Marine Diesel Engines

Marine Diesel Engines

by Daniel P. Charnews

Cornell Maritime Press
Centreville, Maryland

Library of Congress Cataloging-in-Publication Data

Charnews, Daniel, P., 1944-
Marine diesel engines / by Daniel P. Charnews. -- 1st ed.
p. cm.
Includes bibliographical references and index.
ISBN 978-0-87033-590-7
1. Marine diesel motors. I. Title.

VM770.C424 2007
623.87'236--dc22

2007012705

Manufactured in the United States of America
First Edition; first printing, 2007

Table of Contents

Illustrations

Figures

Introduction

This book has been written for a semester course in marine diesel engines. The students taking this course should have a prior understanding of mechanical drawing, introductory thermodynamics and an introductory understanding of diesel engines. The book stresses basic generic principles for all engines through the use of simple, clear explanations and diagrams. Study questions are provided at the end of the chapters to aid the student.

A CD is available from the author (e-mail danchar18@optonline.net). The CD contains high-resolution pictures keyed to the pages in the book to provide detailed examples of the subject matter, a virtual reality model of a slow-speed diesel engine, and a complete jacket water cooling system. This provides a three-dimensional visualization of a diesel engine system that can be explored by the students.

Acknowledgments

I would like to thank Stanley Christensen, Professor Emeritus at the United States Merchant Marine Academy, for his inspiration and guidance in my work at the Academy, and my wife Barbara, for her help in editing this book. The Society of Naval Architects and Marine Engineers and the Institute of Marine Engineering Science and Technology, London, England were of great assistance in my writing. Also, my fellow professors and students at the United States Merchant Marine Academy have been a great help to me and worked hard making the Academy the great learning institution it is today.

Chapter 1
Short History of the Marine Diesel Engine

1629-95

Christiaan Hugens invented a gunpowder engine. This primitive engine used gunpowder to blow out the gases in a cylinder through a check valve. The partial vacuum created pulled a piston into the cylinder. This engine was just a toy.

1824

Nicholás Leónard Sadi Carnot had the idea of compressing air in a cylinder with a compression ratio of fifteen to one and injecting fuel into the cylinder. The air would be hot enough to ignite the fuel. He also suggested some cooling was needed. No one followed this idea up until Herbert Ackroyd-Stuart in 1890.

1860

Jean Joseph Etienne Lenoir developed a double-acting engine, much like a double-acting steam pump. This engine ran on coal gas and was fired by an electric spark. Because it did not have a compression stroke, the efficiency was only 4 percent. Since coal gas was used for gas lighting during this period, the fuel was readily available. The engine package was small because it did not have a big boiler. It required less maintenance, since it did not have a boiler that would foul with soot. Low-pressure steam plants of the day were 12-13 percent efficient, but they were large.

1876

Nicholas Otto used the idea of Beau de Rochas (patent but no engine) to make an engine with an intake, compression, power, and exhaust stroke. This engine looked like a single cylinder engine of today, but it used flame ignition and burned coal gas. It had a 16 percent thermal efficiency. Coal gas was a byproduct of coke. Coke was used to make iron and the coal gas was used in street lamps.

Rudolf Diesel was an engineering student at this time. He knew of the very low efficiency of all these engines. Later he would try to solve this problem.

1873

George Brayton was an American who built an engine that had a separate cylinder for compression and power. It used flame ignition to ignite coal gas, and eventually worked on kerosene. This was a breakthrough, since the fuel source could be carried along with the engines. No one in America wanted the engine, so he moved his operation to England.

1885

Gottlieb Daimler reworked the Otto engine to run on gasoline. He got together with another German and started an automobile company, Daimler–Benz.

1890

Herbert Ackroyd-Stuart was an Englishman who had an engine very similar to a modern diesel engine. It had a precombustion chamber with a copper hot spot that was heated with a blowtorch, for starting. This engine also used solid injection and ran on kerosene.

1892

Rudolf Diesel obtained his first patent. The engine ran on coal dust with a 1,500 psi compression pressure and no cooling. The idea was to have such an efficient engine that the expansion would take care of the cooling. The engine blew up and almost killed Diesel.

1895

Diesel tried again with an engine that had a 450 psi compression pressure, was water cooled, had air blast injection, and ran on fuel oil. This engine ran at 24 percent efficiency, which was very good at the time. However, Diesel wanted to build an engine that approached the theoretical efficiency of his first engine, which blew up. He never accomplished this feat.

As you can see, Diesel was not the only person that had developed an internal combustion engine that was ignited by compression heat, but history has given him most of the credit. As far as the efficiency goes, engineers have only recently designed engines that approached the efficiency Diesel wanted his engine to have. Today the slow-speed, turbocharged, two-stroke cycle engine runs at a little more than 50 percent efficiency. This is the engine efficiency itself. Things like a waste heat boiler and cooling water to run the evaporator will use even more of the engine heat that would have been wasted.

1910

The first diesel-powered ship was the tanker *Vulcanus,* commissioned by Anglo Saxon Shell.

1912

The diesel ship *Selandia* was commissioned.

As an example demonstrating the fuel efficiency of the diesel engine, several ships have been converted from steam to diesel. The *Mobile Hawk* burned 105 tons of fuel a day as a steam ship. After conversion to diesel, the same ship burned thirty-eight tons of fuel a day. As you can see, that is a big difference, especially considering the change in crude oil prices. That is why virtually all new ships are being powered by diesel engines.

Chapter 2
Thermodynamic Review

Boyle's Law

If the temperature is constant, the product of pressure and volume of an ideal gas is a constant or the pressure varies inversely as the volume.

$$PV = \text{Constant}$$
$$P_1V_1 = P_2V_2 = \text{Constant}$$

Charles's Law

If an ideal gas is heated at constant pressure, the volume divided by the temperature is a constant or the volume varies directly as the temperature.

$$\frac{V}{T} = \text{Constant}$$
$$\frac{V_1}{T_1} = \frac{V_2}{T_2} = \text{Constant}$$

Universal Gas Law

These laws can be combined as follows:

$$\frac{PV}{T} = \text{Constant}$$
$$\frac{P_1V_1}{T_1} = \frac{P_2V_2}{T_2} = \text{Constant}$$

The constant in this law is made up of the mass of the gas and the gas constant. The units of the gas constant dictate the units of the other terms.

$$\frac{PV}{T} = mR$$
$$PV = mRT$$

If we do not know the mass of the gas we are working with, we can use one unit of mass. This means we use the specific volume as shown below.

$$Pv = RT$$
$$v = \frac{V}{m}$$

Another way to write these laws and introduce a new equation is shown below.

$$PV^{n} = \text{Constant}$$
$$P_1V_1^{n} = P_2V_2^{n} = \text{Constant}$$

n changes depending upon the process. If n=1 we have Boyle's Law. When working with diesel engines, we need to use a reversible adiabatic or constant entropy process for compression and expansion. In this case, n would be equal to the ratio of the specific heats of the gas, which will name k.

$$n = k = \frac{C_P}{C_V}$$

This is only for an isentropic or reversible adiabatic process. For a real process, this value can be modified.

You must remember that the temperature and pressure must be in absolute units for these equations to work. For example, if you started with the following at bottom dead center for the compression process of a diesel engine:

Air temperature: 90 °F

Pressure: 1 atm absolute n=k=1.4 (Air)

Compression ratio: 22/1

If you choose the correct equations, you would find at top dead center using an isentropic compression process the following:

Pressure at top dead center: 1099 psig

Temperature at top dead center: 1434 °F

You also have to know the compression ratio is a volume ratio. This compression ration equals the cylinder volume at bottom dead center divided by cylinder volume at top dead center.

Temperature Scale Conversion Formulas

$$R^o = F^o + 460$$

$$K^o = C^o + 273$$

$$F^o = \frac{5}{9}C^o + 32$$

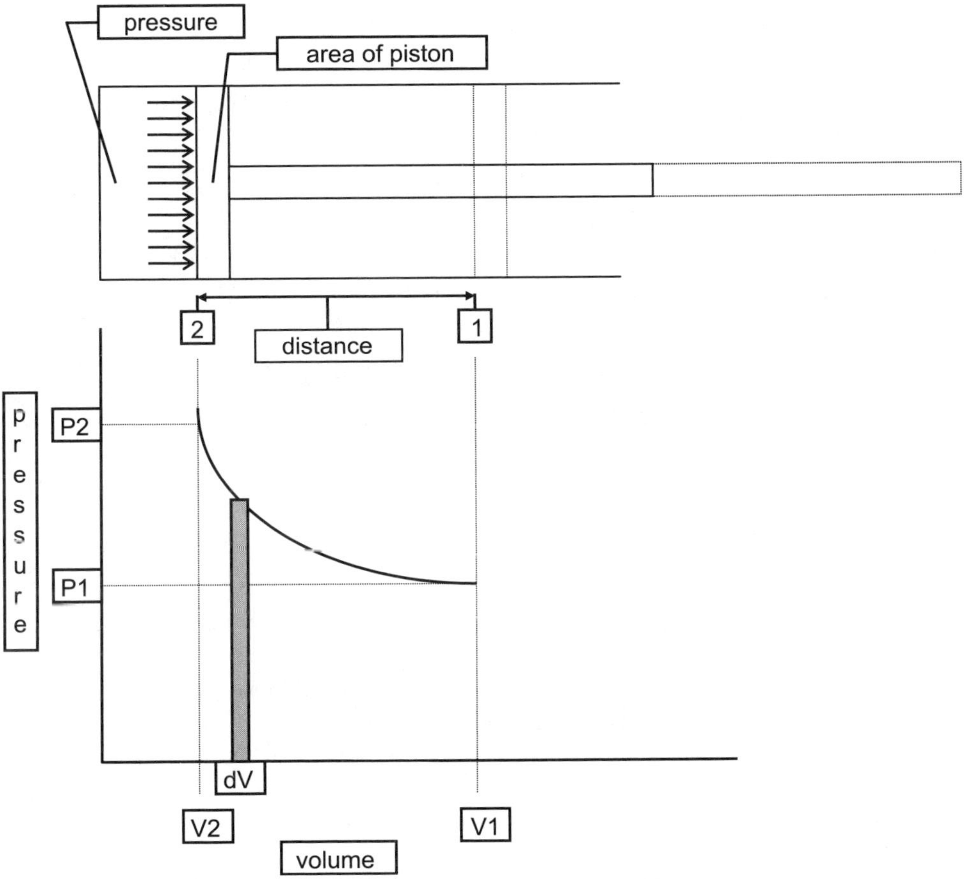

Fig. 2.1. Pressure volume diagram

Fig. 2.1 represents an adiabatic compression from point one to two. To do this compression process we would have to put work into the system. The definition of work is a force times a distance. Force and distance are vectors. They must line up. So this multiplication must be a dot product. Luckily for us, in a diesel engine the force and the distance line up because the piston is fitted into the cylinder. Thus, a simple product will do.

$$\text{Work} = \text{Force} \circ \text{Distance}$$
$$W = \vec{F} \circ \vec{D}$$

Since the force is lined up with the distance, we have the following:

$$W = FD$$

The force in our case is equal to the pressure times the area of the piston.

$$\text{Force} = \text{Pressure} \otimes \text{Area}$$
$$F = PA$$

Substituting the force expression into the work equation.

$$\text{Work} = \text{Pressure} \otimes \text{Area} \otimes \text{distance}$$
$$W = PAd$$

However, Ad is the change in Volume going from one to two.

$$W = P\Delta V$$
$$dW = PdV \quad \text{as Delta V goes to zero}$$

This means that the area under the curve in fig. 2.1 is equal to the work. If you calculate the units of P Delta V, you will get foot pounds. Whether the work is going in (-) or coming out (+) of the system, depends on whether you are compressing the gas on top of the piston (-) or the gas is expanding (+).

We can use this information to find the net work of the cycle. First, let us look at some very simple theoretical cycles. Actual cycles are very complex, with gases flowing in and out of the engine. We will start with air standard cycles that can be modeled easily, using the gas laws we have looked at.

Air Standard Cycle

1. The working substance is air and it stays in the cylinder at all times.
2. Heat is added instead of fuel being burned.
3. Heat is rejected instead of exhaust being expelled.

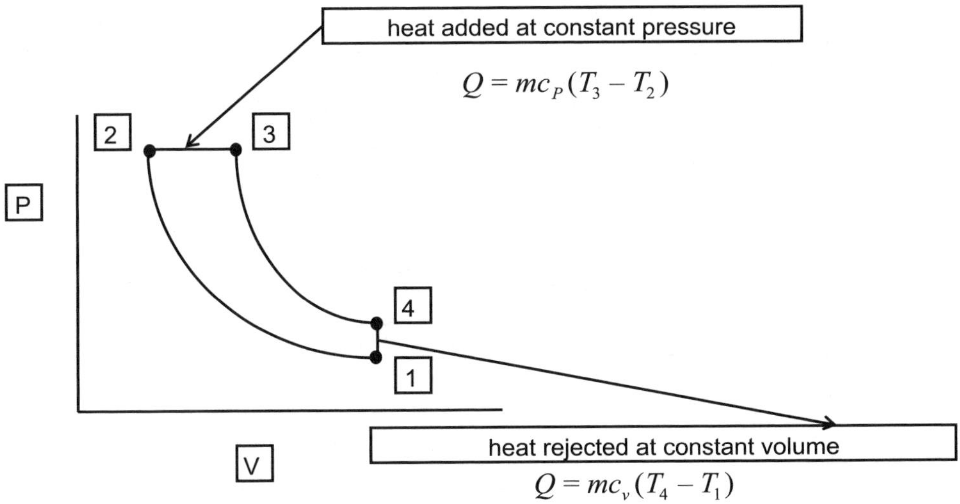

Fig. 2.2. Air standard diesel cycle

One to two Adiabatic compression with no heat transfer across cylinder boundary
Two to three Heat addition at constant pressure
Three to four Adiabatic expansion no heat given off
Four to one Heat rejected at constant volume

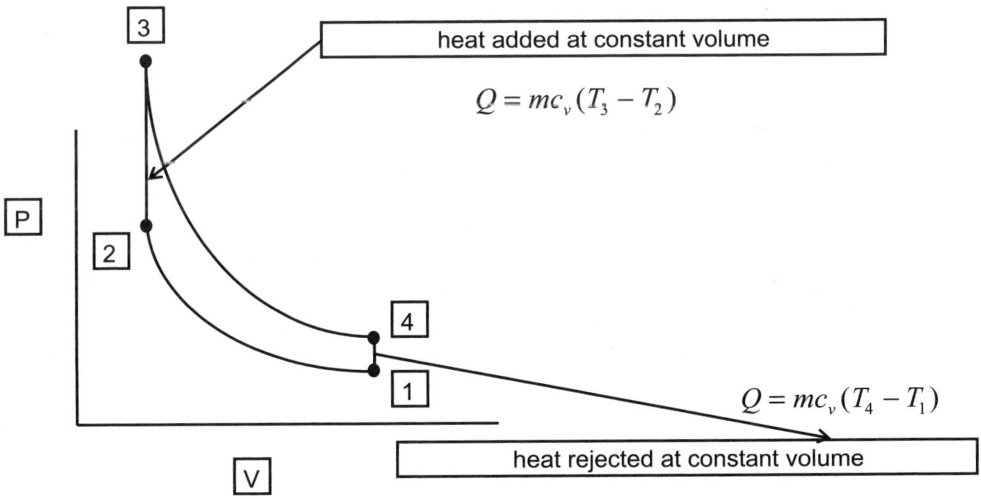

Fig. 2.3. Air standard Otto cycle

The basic difference between the Otto cycle and the Diesel cycle is in the addition of heat (fig. 2.3). The Otto cycle models the gasoline engine, adding heat at constant volume. The Diesel cycle theoretically adds heat at constant pressure.

An actual Diesel cycle adds heat at both constant volume and constant pressure. To model this, the air standard dual cycle is used.

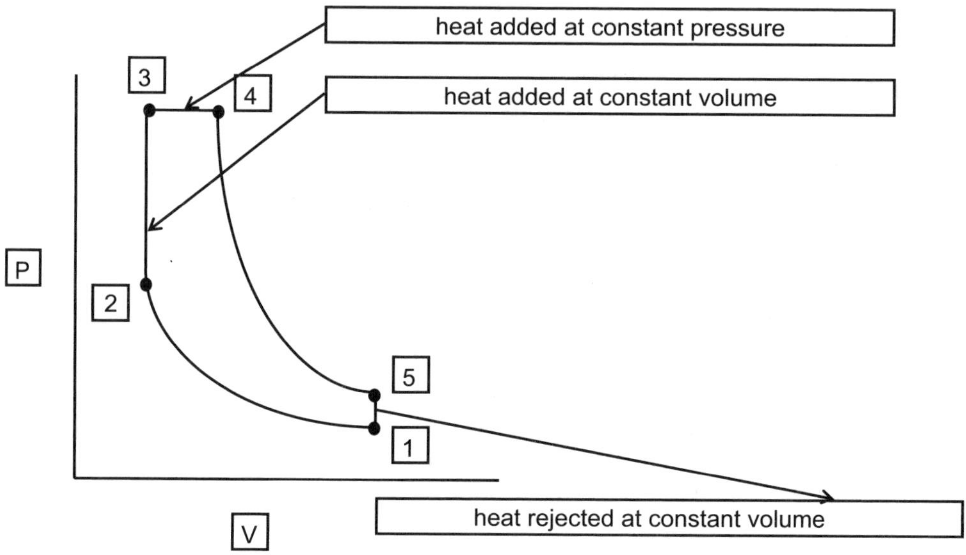

Fig. 2.4. Air standard dual cycle

To see how these theoretical cycles are used, let us try an example. To start with, we must choose some reasonable initial conditions for point one. We might as well choose the most complicated cycle for the example, which would be the dual cycle.

Since most diesel engines are turbocharged, a reasonable pressure for point one would be two atmospheres gauge pressure. Since we would have an intercooler, the temperature for point one would be 100 °F. A reasonable compression ratio would be 15/1. This is fairly low for a naturally aspirated engine, but the turbocharger causes a high pressure at the start of compression.

Using the following equations, we can start at point one and go around the dual cycle and find the temperature, pressure and volume for the rest of the points.

$$Q = mc_P(T_4 - T_3)$$

$$Q = mc_v(T_3 - T_2)$$

$$n = k = \frac{C_P}{C_V}$$

$$PV^n = \text{Constant}$$

$$P_1V_1^n = P_2V_2^n = \text{Constant}$$

$$\frac{PV}{T} = \text{Constant}$$

$$\frac{P_1V_1}{T_1} = \frac{P_2V_2}{T_2} = \text{Constant}$$

$$Pv = RT$$

$$v = \frac{V}{m}$$

Going from point one to point two:

Using $\frac{P_1V_1}{T_1} = \frac{P_2V_2}{T_2}$ and with $P_1V_1^n = P_2V_2^n$ k=1.4 for an adiabatic compression of air we would find:

$$P_2 = 132.9 \quad \text{atm abs}$$

$$T_2 = 1653^oR$$

Going from point two to point three:

Using $Q = mc_v(T_3 - T_2)$ with $c_v = .18 Btu/lb^oR,$ m=1 lb of air and choosing to add 100 Btu to the cylinder at constant volume, we would find:

$$T_3 = 2208^oR$$

For the dual cycle:

Using $\frac{P_3V_3}{T_3} = \frac{P_2V_2}{T_2}$ with the volume being constant we would find:

$$P_3 = 177.5 \quad \text{atm abs}$$

Going from point three to point four:

For the dual cycle:

Using $Q = mc_P(T_4 - T_3)$ with $c_P = .25 Btu/lb^oR,$ m=1 lb of air and choosing to add 100 Btu to the cylinder at constant pressure we would find:

$$T_4 = 2608^oR$$

Using $Pv = RT$ with R=53.3 ft-lb/lb °R remembering this gas constant dictates the units, thus pressure must be in pounds per square foot absolute. Since we are using one pound of air in this example, we effectively are using the specific volume for the volume. Thus, we can find the specific volume at point 4 or at any other point as long as we know the temperature and pressure. Since we are going from point 3 to point 4 at constant pressure $P_3 = P_4$

Thus, $$v_4 = .37 ft^3 / lb$$

Going from point four to point five:

Since $v_5 = v_1$ and $Pv = RT$ (remember P lb / ft^2) we can find $v_1 = 4.7 ft^3 / lb$

Using $P_5 v_5^n = P_4 v_4^n$ and adiabatic expansion we can find:

$$P_5 = 5 \text{ in atm abs}$$

Using $\dfrac{P_4 V_4}{T_4} = \dfrac{P_5 V_5}{T_5}$ we can find $T_5 = 933^oR$

Now we have all the information on all the points we need. We can now calculate how much heat is exhausted.

Using $Q = mc_v(T_5 - T_1)$ we can find $Q_{exh} = 67 Btu$

with $Work_{out} = Q_{in} - Q_{out}$ = 133 Btu every time we go around the cycle.

Efficiency = (Work out/Heat in)X 100% = 66.5%

We have figured out the efficiency of the dual cycle with the parameters we have chosen.

We could go on and answer more questions about an engine using the above information. For example:

1. We could use the heating value of a particular fuel and covert the Qin of 200 Btu into an amount of fuel.
2. We could use an air-fuel ratio to check and see if we have enough air to burn the fuel.
3. We could calculate the output of the engine, using the work out, rpm, number of cylinders, and type of cycle. This would give the output in Btu per minute and we could convert this to horsepower.
4. We could use the specific volume to find the engine displacement.

More accurate results can be obtained by using air tables because air is not an ideal gas.

The efficiency of the previous cycles can be calculated in general terms. The efficiency can then be plotted versus a number of variables. I have chosen to plot efficiency versus compression ratio as shown in fig. 2.5.

As you can see, the Otto cycle is the most efficient and the Diesel cycle is the least efficient. However, you probably know that the diesel engine is the most efficient, while the gasoline engine is not as efficient. The Otto cycle most closely models the gasoline engine.

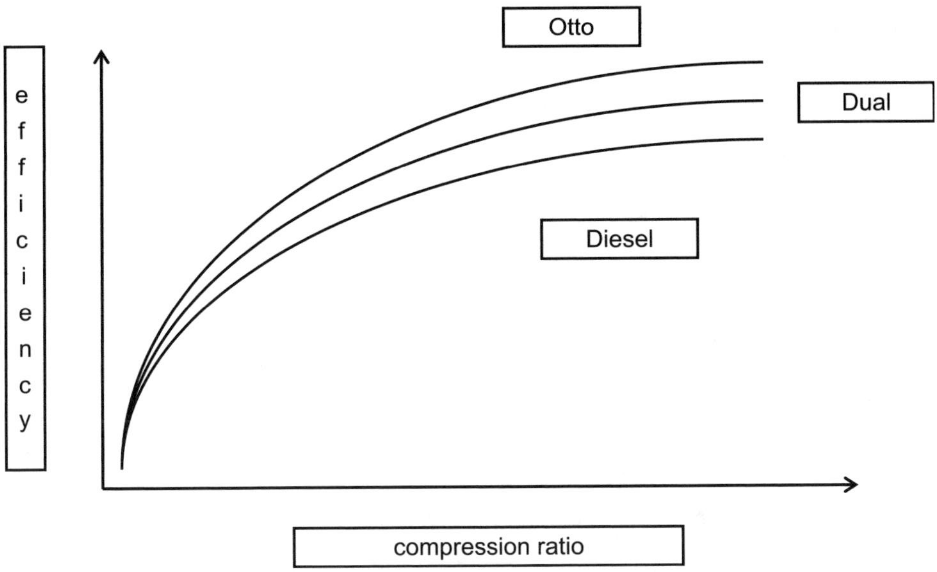

Fig. 2.5. Efficiency versus compression ratio

This makes sense if you take into account the engines run at different compression ratios. The actual diesel engine runs at higher compression ratios than the gasoline engine. That is why the actual diesel engine is more efficient.

What makes the Otto cycle so efficient? It is because all the energy used is introduced at top dead center. This means all the energy is expanded to bottom dead center. In all the other cycles some of the energy is introduced after top dead center and this energy cannot push on the piston throughout its stroke.

Study Questions

1. Which of the following men have an air standard cycle named after them?
   A. Dual
   B. Otto
   C. Carnot
   D. Daimler
2. If an ideal gas is heated at constant pressure the volume varies __________ as the temperature.
   A. directly
   B. the same
   C. inversely
   D. exactly
3. Start with the following at bottom, dead center: Air temperature 90 °F; Pressure 1 atm absolute Cp/Cv=1.4 (Air); Compression ratio 22/1

   What would be the pressure at top dead center?
   A. 22 atm abs
   B. 1,434 psia
   C. 1,099 psia
   D. 1,099 psig
4. The area under a curve represents the work performed in a process in which graph?
   A. PV
   B. PT
   C. PE
   D. Ts
5. Which process represents heat being added in diagram A?
   A. 7
   B. 8
   C. 9
   D. 10

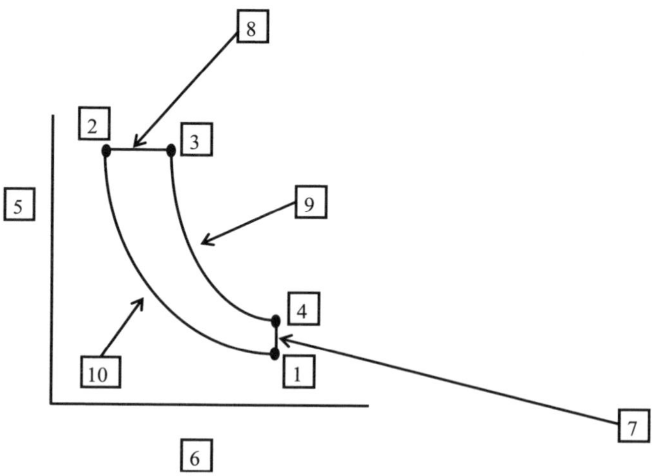

Diagram A

6. What process in diagram A would represent fuel being injected in a real engine?
   A. 7
   B. 8
   C. 9
   D. 10
7. Which air standard cycle does diagram A represent?
   A. Diesel
   B. Dual
   C. Otto
   D. Carnot
8. Which process represents the heat being rejected in diagram A?
   A. 7
   B. 8
   C. 9
   D. 10
9. What does 5 represents in diagram A?
   A. pressure
   B. temperature
   C. volume
   D. entropy
10. What is the most efficient air standard cycle for a given compression ratio?
    A. Diesel
    B. Dual
    C. Otto
    D. Carnot
11. If the temperature is kept constant the pressure of an ideal gas varies ____________ as the volume.
    A. directly
    B. the same
    C. inversely
    D. exactly
12. Which term in the following expression dictates the units that must be used in all the other terms? PV= mRT
    A. P
    B. V
    C. R
    D. T
13. What is the most efficient working engine?
    A. diesel
    B. gasoline
    C. steam turbine
    D. gas turbine

Chapter 3
Practical Cycles

Chapter two examined theoretical I.C. engine cycle pressure volume diagrams. Pressure volume diagrams can be taken from an actual engine. The device that does this is called an indicator (fig. 3.1) Since this device is mechanical, it can only operate at low engine revolutions, and we can only take indicator cards of slow-speed engines.

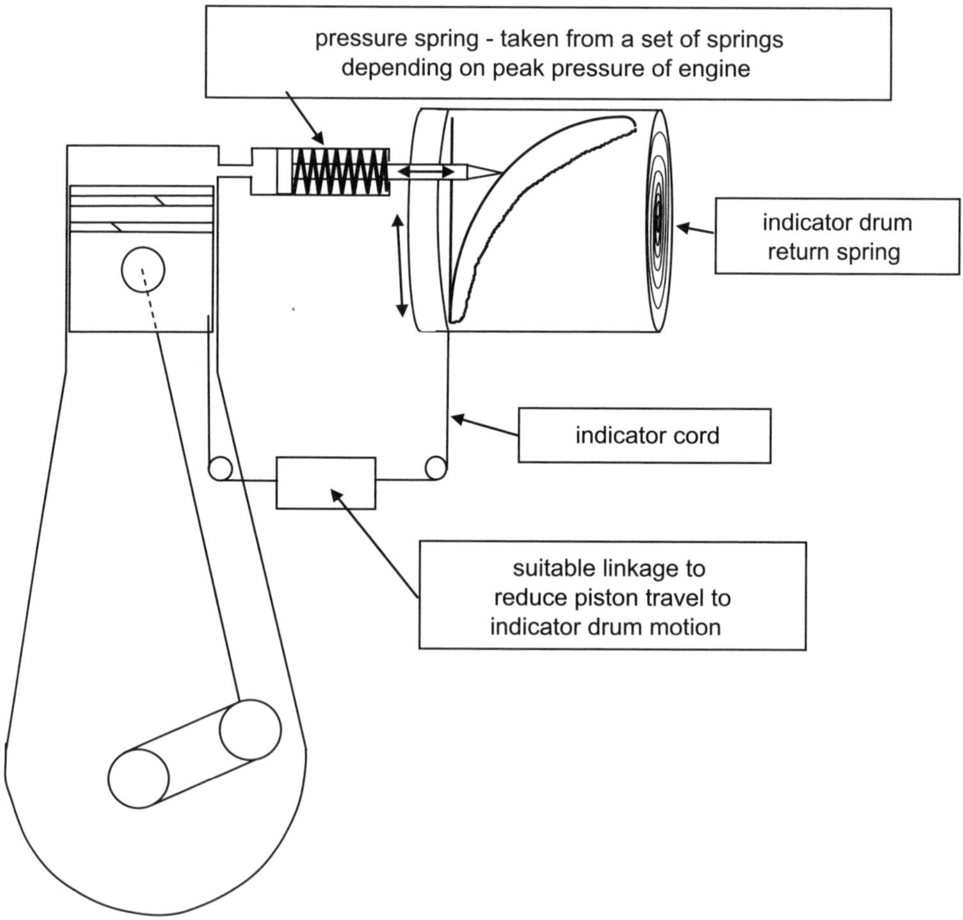

Fig. 3.1. Simple diagram of indicator mechanism

The indicator is a mechanical device that obtains the pressure-volume diagram from a slow-speed engine. The piston and pressure spring arrangement senses the pressure. The cord pulling on the drum registers volume.

If we want to take pressure volume diagrams of medium- and high-speed engines, we have to do it electronically. This can be done with a piezoelectric pressure transducer, charge amplifier, and oscilloscope. A computer with a data acquisition board usually replaces the oscilloscope today.

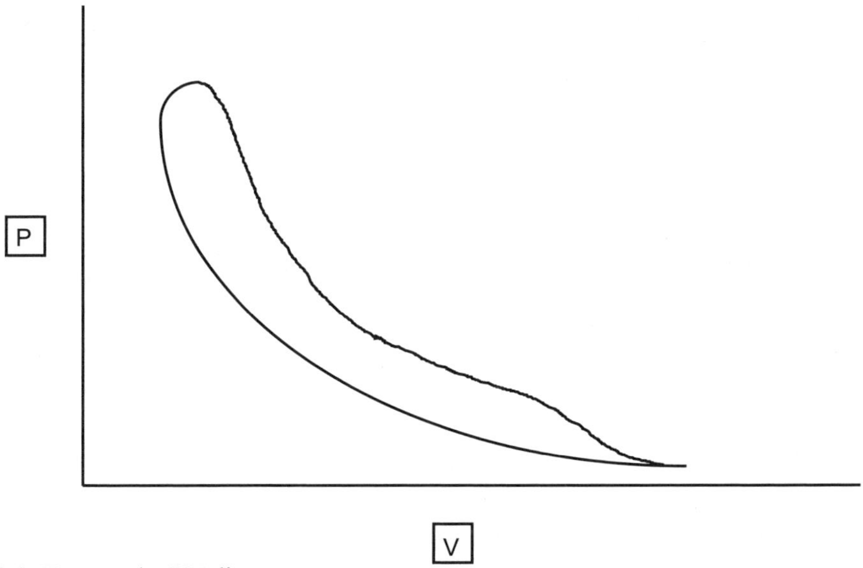

Fig. 3.2. Two-stroke PV diagram

The PV diagram shown in fig. 3.2 would be from a two-stroke cycle engine.

As you can see, it does not exactly look like any of the theoretical cycles, but it resembles the Dual cycle.

To use the actual PV diagram, we can use the fact that the area inside the trace represents the work that the cycle outputs each time it is executed. The problem we have in determining the amount of work is that we have such an irregular area. The problem can be solved using a planimeter. We can determine the area inside the trace with the planimeter in square inches. We can then draw an equivalent rectangle on the PV diagram fig. 3.3 using the same length as the length of the trace and the height equal to the trace area divided by the trace length. We are working in two domains: The picture domain of inches and square inches with the planimeter and the PV diagram domain of pressure, volume, and work. It is a simple matter to find the area of this rectangle using the units of pressure and volume in the PV domain. If you multiply the height of the rectangle in the pressure units of pounds per square foot by the length of the rectangle in volume units of cubic feet you will get work in the units of foot pounds.

This height of the rectangle in units of pressure has the specific name of mean indicated pressure or mean effective pressure.

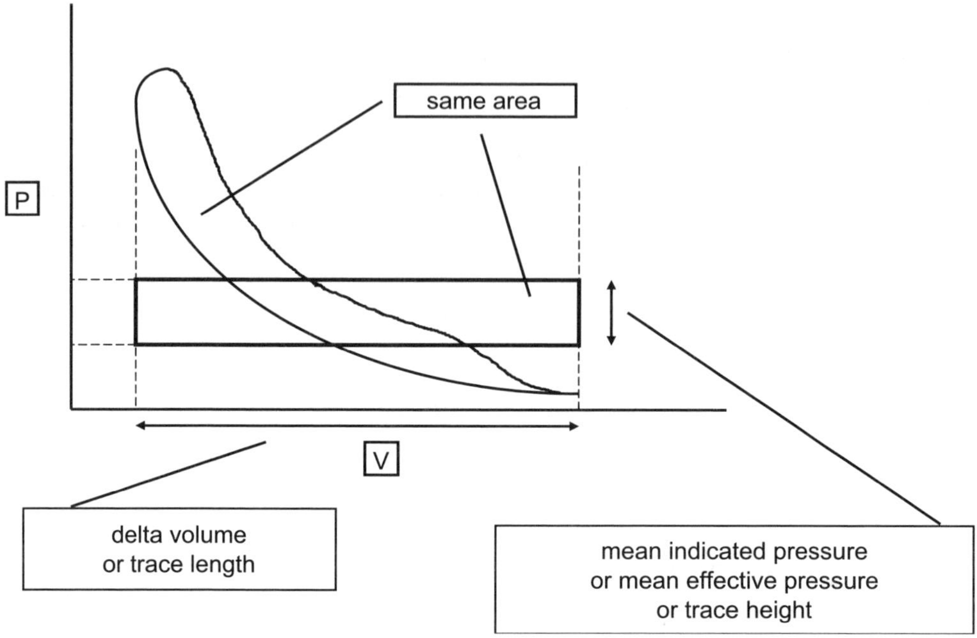

Fig. 3.3. PV diagram equivalent areas

This mean effective pressure usually is in the units of $\frac{lb}{in^2}$. If you multiply the height of the rectangle in inches in the picture domain by the vertical scale factor in psi/in, you get the average pressure in psi in the PV domain.

This method only works if we can obtain a pressure versus volume trace and use the planimeter. For medium- and high-speed engines, we probably would use a computer. A mechanical indicator has too much inertia to follow the motion in these engines. The computer gives us a pressure time trace instead of a pressure volume trace.

This is the pressure versus time trace of a four-stroke cycle diesel engine (fig. 3.4). Note the two traces acquired by the computer. We need both traces to find the mean effective pressure. We need the top dead center (TDC) tick to find the individual strokes that make up the cycle. Note that bottom dead center (BDC) is directly between the TDC ticks.

We could construct a PV diagram from this diagram by pulling out the individual strokes from the pressure trace and plotting them, starting at TDC and going to BDC. We must remember to reverse the compression and exhaust strokes before plotting them. This is because they start at BDC and end at TDC. We then could use the planimeter to find the mean effective pressure.

However, since we have the trace stored in a computer, we can use software to calculate the mean of the stroke. The scale factor of psi/volt is needed. We must find the mean of each stroke because they represent the area under the stroke. Some of this area is output work that turns the flywheel. Some of the area is input work that compresses air. That is why the area inside the cycle represents the net work. In this case, the major part of the net

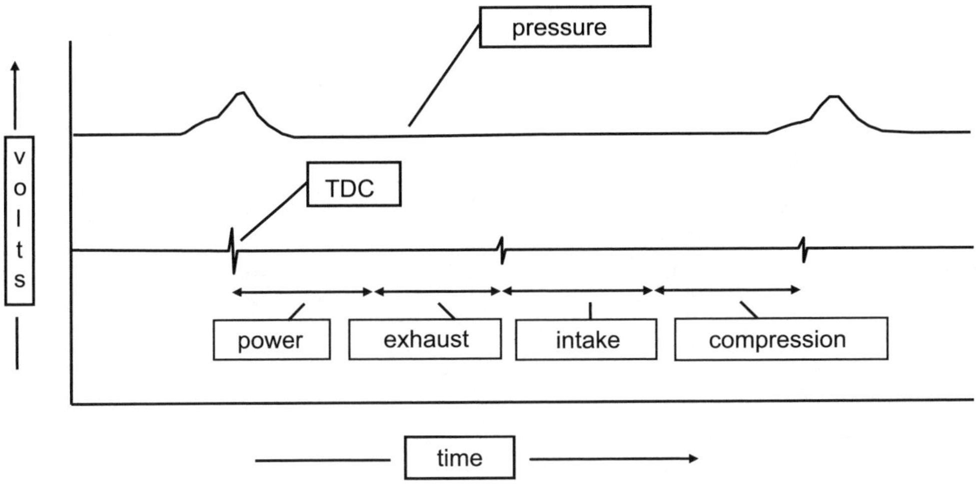

Fig. 3.4. Pressure vs. time trace of a four-stroke cycle engine

work out is represented by the mean of the power stroke, minus the mean of the compression stroke. I will call this the net mean pressure of the power and compression strokes.

The intake and the exhaust stroke also contribute a smaller amount of work. In a naturally aspirated engine the mean of the exhaust will be higher than the mean of the intake. This means that the engine is pumping exhaust out and intake in. So this work is not net work out for these two strokes. They are work going into the engine. You subtract the mean of the intake from the mean of the exhaust, then subtract this value from the net mean pressure of the power and compression strokes.

In a turbocharged engine, the intake is higher than the exhaust. This is because the air compressor part of the turbocharger is pumping in air under pressure and actually pushing down on the piston to turn the crankshaft. We subtract the mean of the exhaust from the mean of the intake and add this value to the net mean pressure of the power and compression strokes.

With the electronic system, the pressures are read in volts, so you need the conversion factor of psi/volt to get the mean effective pressure in psi.

This is another reason why turbocharging an engine increases efficiency. Not only does the turbocharger compress the intake air before the engine has to compress this air, but it also pushes down on the piston during the intake stroke to help turn the crankshaft. This is all done by using exhaust gas energy that otherwise would have been wasted.

If we construct the PV diagram by computer or we used an indicator, the diagram for a four-stroke cycle engine would look like fig. 3.5.

In the scale of the power and compression strokes, the intake and exhaust appear as a flat line. If we wanted more detail in this area, we have to use software to zoom in on the intake and exhaust strokes. We also could use a weaker pressure spring in our mechanical indicator to expand this area. This is known as a weak spring indicator diagram.

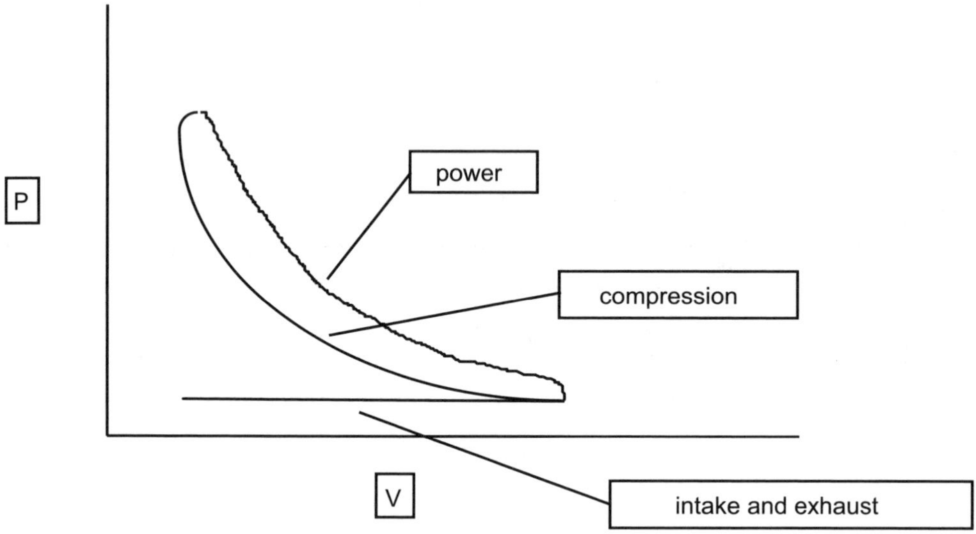

Fig. 3.5. Four-stroke PV diagram

Power Equations

Now that we have the mean effective pressure (MEP), or what we will call (IMEP), we can use this parameter to calculate the indicated horsepower (IHP). The well-known formula for IHP can be derived from basic concepts.

$$\text{Work} = \text{Force} \circ \text{Distance}$$

$$W = \vec{F} \circ \vec{D}$$

Since the force is lined up with the distance, we have the following:

$$W = FD$$

The force in our case is equal to pressure times area:

$$\text{Force} = \text{Pressure} \otimes \mathit{Area}$$

$$F = PA$$

Substituting the force expression into the work equation:

$$\text{Work} = \text{Pressure} \otimes \text{Area} \otimes \text{distance}$$

$$W = PAd$$

In this case, we will use the (MEP or IMEP) mean effective pressure or indicated mean effective pressure for P.

A will be the cylinder cross-sectional area.

L will be the length of the stroke, the distance d the piston travels.

$$W = PLA$$

If the units of P are pounds per square inch, and the units of

A are square inches, and the units of

L are feet,

the work will be in units of foot-pounds.

This means every time we complete a cycle, we output this amount of work.

We now have to incorporate time into this equation. Power is work divided by time.

One horsepower is 33,000 (ft-lb)/min.

The time variable has to incorporate the rpm, number of cylinders, and type of cycle of the engine.

N = rpm x number of cylinders x 1 or 1/2

1 is used for the two-stroke cycle

1/2 is used for the four-stroke cycle

This is because it takes one revolution to complete the two-stroke cycle and two revolutions to complete the four-stroke cycle.

N represent the number of cycles per minute for the whole engine.

Putting this all together, we have the classic expression for indicated horsepower (IHP).

$$IHP = \frac{PLAN}{33000}$$

Remember the units are dictated by the 33,000 foot pounds/minute horsepower, conversion factor.

A six-cylinder, four-stroke cycle engine runs at 3,000 rpm, with a mean indicated pressure of 200 psi. This engine has a bore of 4 inches and a stroke of 6 inches. The IHP of this engine would be about 343 horsepower.

While we are talking about horsepower, there is another term referred to as brake horsepower. This power originally was measured at the output shaft of the engine using a brake. Today we can use a water brake or dynometer to measure this output power.

This leads us to a new pressure definition called the brake mean effective pressure (BMEP).

Brake Mean Effective Pressure (BMEP)

We calculate the BMEP by using the measured brake horsepower. We then take the brake horsepower and the engine particulars and use them in our equation to solve for P, which will be the BMEP. The BMEP is always lower than the indicated mean effective pressure, due to the internal friction of the engine. Of course, using HP=PLAN/33000 for the same engine, the brake horsepower is always less than the indicated horsepower.

The goal of the engine designer is to use these tools to design the most powerful and efficient engine. This can be done using the most efficient cycle and the highest mean effective pressures. How high a mean effective pressure we can use is limited, based on the peak pressure in the engine. Too high a peak pressure can damage the engine due to the high pressure and the associated high temperature. Thus, the peak and mean effective pressures are limited by the current materials used to construct diesel engines.

As we have seen before, turbocharging increases the efficiency and power output of the engine. Again, maximum pressures and temperatures have to be watched.

The type of cycle has two aspects. The first is a two-stroke cycle or four-stroke cycle. You would think that the two-stroke cycle would be twice as powerful as a four-stroke cycle, but the two-stroke cycle has breathing problems. The four-stroke cycle has two whole strokes to breathe. Firing every revolution puts a large heat load on two-stroke cycle engines, limiting the amount of fuel fed to the engine. This limits power output. With modern engines and turbocharging, the operating cycles are about equal as far as power and efficiency.

The second cycle aspect is the theoretical cycle. We are mainly concerned with diesel engines operating on the theoretical Diesel cycle. However, if we could operate this diesel engine on the Otto cycle, we would have a more efficient engine. We can do this by injecting the fuel as fast as possible to simulate the heat added at constant volume of the Otto cycle, as seen in fig. 3.6. Thus, all the energy can expand from TDC to BDC.

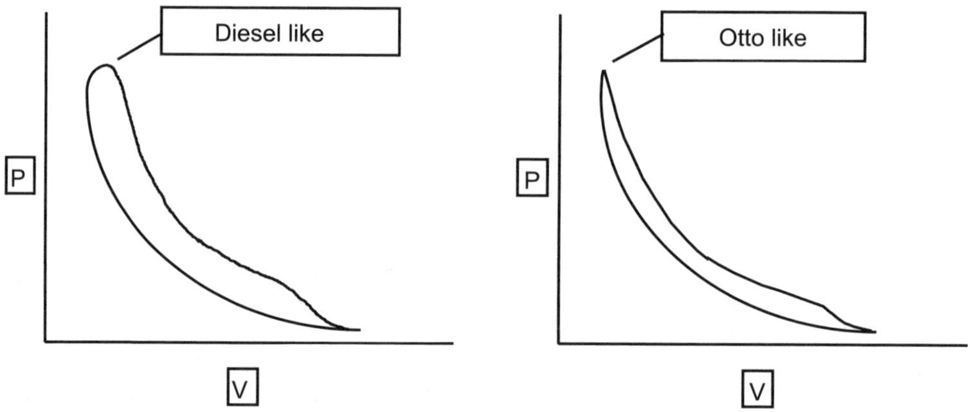

Fig. 3.6. Two-stoke PV diagram comparison

Notice that the above PV diagrams are of two-stroke cycle slow-speed engines. This is the only engine available that has the choice of tuning the fuel system to operate as a Diesel or Otto cycle engine. Note that the peak pressures are the same because of the possibility of damaging the engine. The work output of the Otto cycle engine has been reduced, as seen by the smaller area enclosed by the trace, because we have to keep the same peak pressure. We can either have a less powerful, efficient engine or a more powerful, less efficient engine for the same displacement.

The most efficient engine available today is a turbocharged, two-stroke cycle, slow-speed diesel engine tuned to run as close as possible to an Otto cycle.

Study Questions

1. What does diagram B represent?
   A. air standard Diesel cycle
   B. air standard Otto cycle
   C. actual four-stroke diesel cycle
   D. actual two-stroke diesel cycle
2. What does the gray area in diagram B represent?
   A. net work out
   B. net work in
   C. work of compression
   D. work of expansion

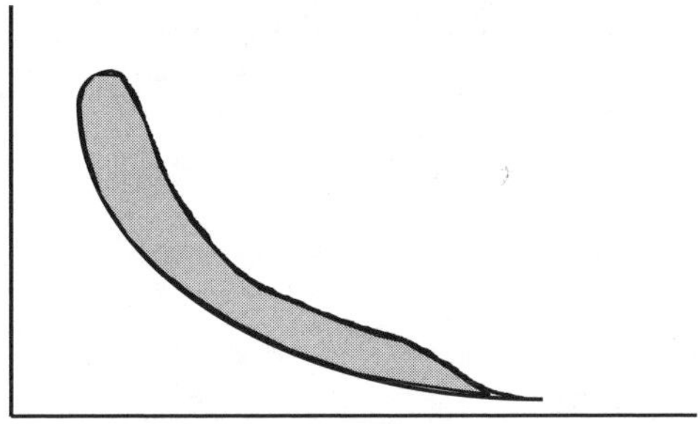

Diagram B

3. In diagram C, arrow 2 represents:
   A. clearance volume
   B. average pressure
   C. mean indicated pressure
   D. mean indicated volume
4. The units of the horizontal axes of diagram C would be:
   A. cubic feet
   B. square feet
   C. feet
   D. degrees Rankine
5. In diagram C, the area enclosed by the kidney shape is:
   A. greater than the area enclosed by the dashed rectangle
   B. the same as the area enclosed by the dashed rectangle
   C. less than the area enclosed by the dashed rectangle
   D. is not related to the area enclosed by the dashed rectangle
6. In diagram C, arrow 1 represents:
   A. total cylinder volume
   B. cylinder pressure
   C. length of the connecting rod
   D. length of the indicator diagram

7. Diagram C has to be obtained by what device in a high-speed engine?
   A. indicator
   B. planimeter
   C. protractor
   D. electronic engine analyzer

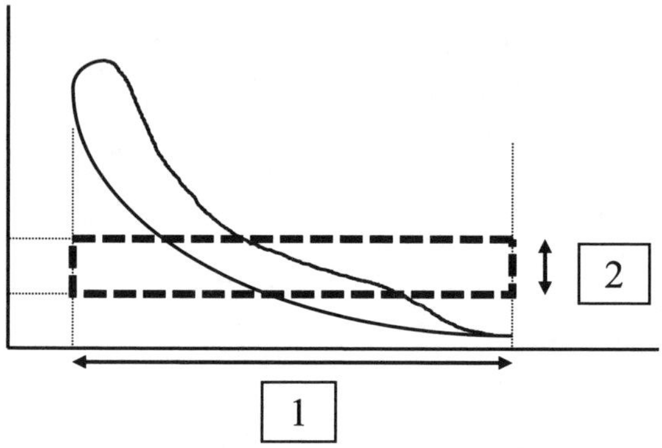

Diagram C

8. In formula A, the letter A represents:
   A. area of the indicator card
   B. cylinder bore area
   C. inlet air area
   D. exhaust area

$$IHP = \frac{PLAN}{33000}$$

Formula A

9. In formula A, the letter P represents:
   A. plane area
   B. mean effective pressure
   C. peak pressure
   D. peak area

10. A six-cylinder, two-stroke cycle engine is running at 3,000 rpm with a mean indicated pressure of 200 psi. This engine has a cylinder bore of four inches and a stroke of six inches. What is the Indicated Horsepower?
    A. 686
    B. 343
    C. 586
    D. 443

Chapter 4

Engine Selection, Performance, and Fuel Rate Calculations

Engine Selection

The most efficient engine is the turbocharged, slow-speed, two-stroke-cycle diesel engine. A variety of diesel engines are available for different applications. These can be summarized in fig. 4.1.

The governing formula for size and weight verses rpm is the brake horsepower formula.

$$BHP = PLAN/33000$$

The maximum mean effective pressure must remain the same to prevent engine damage. Therefore, for the same horsepower if the rpm goes up, the length of stroke and cylinder area must go down. This means you would have a smaller, lighter engine.

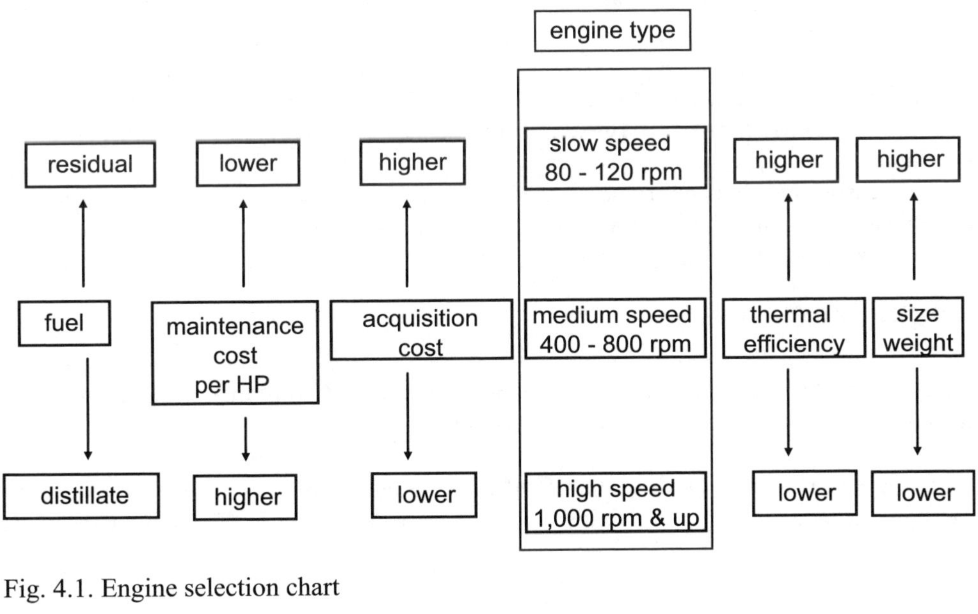

Fig. 4.1. Engine selection chart

Brake-specific fuel consumption (BSFC) is another term used to describe how efficient an engine is.

$$bsfc = \frac{\frac{lb_{fuel}}{hr}}{hp} \text{ If you clear the fraction } \quad bsfc = \frac{lb_{fuel}}{hphr}$$

The fraction usually is cleared when you see the units presented. However, it makes more sense to realize that the BSFC is the pounds of fuel burned per hour to produce one horsepower. Thus, if you know the horsepower of an engine and its BSFC, you can multiply these two values and obtain the amount of fuel this engine burns per hour at this horsepower.

The metric system uses grams of fuel burned per hour to produce one horsepower or one kilowatt.

$$bsfc = \frac{g\ fuel}{hphr} \quad \text{or} \quad \frac{g\ fuel}{kwhr}$$

We now can choose an engine for a particular application based on horsepower, size, and weight. This gives us a high-, medium- or slow-speed engine. We then search the manufacturer's brochures to pick the engine with the lowest BSFC. We also want to talk to operators who have used these engines and make sure the engines are reliable and have low maintenance costs.

The next thing we must do is decide at what load we want to run the engine. We have to know two definitions to discuss this matter.

Maximum Continuous Rating: This is the maximum horsepower and rpm that the engine manufacturer has decided this engine can be run and still have a reasonable life and maintenance costs.

Normal Continuous Rating: This is the normal horsepower and rpm that the owner has decided to run the engine. This value is slightly lower than the maximum continuous rating. Usually the NCR is about 10 percent lower than the maximum continuous rating in both horsepower and rpm.

The reason for this is to promote longer engine life, lower fuel consumption, and less maintenance cost. In other words, an engine run at a slightly lower load and rpm, incurs less wear and tear and burns less fuel.

Engine Performance

To aid in trying to choose at what power output and rpm to run an engine, the manufacturers usually supply a performance diagram. These vary from manufacturer to manufacturer, but the one in fig. 4.2 illustrates best the basic principles that underlie all these diagrams.

What we have here is a diagram of BMEP verses engine rpm. In this way, the diagram can cover a range of engines. To find horsepower, you would use the engine particulars and PLAN/33000.

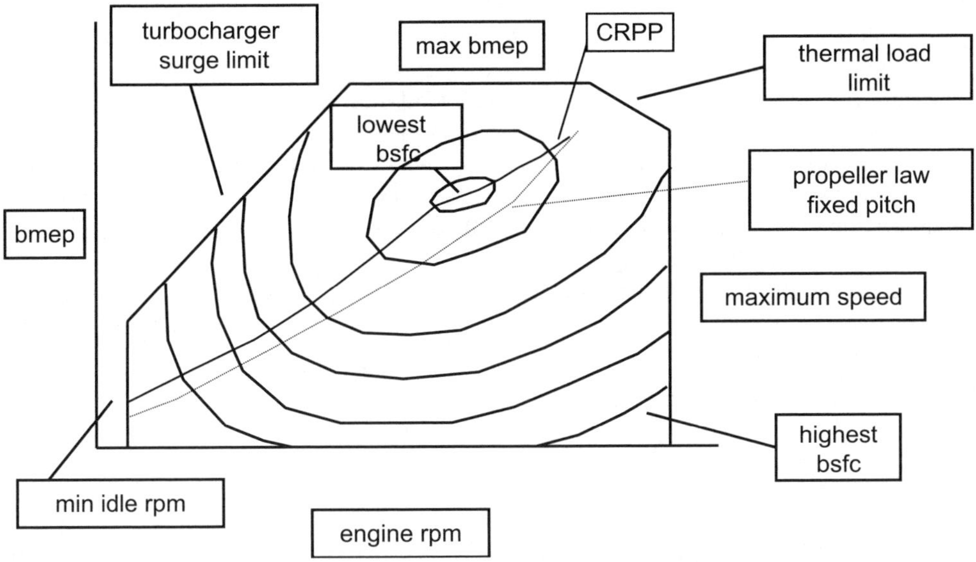

Fig. 4.2. Engine performance diagram

We can run the engine anywhere inside the envelope. However, if the operating point nears any of the boundaries the engine may run into trouble because of the various reasons labeled on the diagram.

A minimum idle speed must be maintained, because if the engine runs at too low a speed, enough air will leak by the rings preventing the engine from firing.

The maximum speed cannot be exceeded, because if the engine runs too fast the inertia loads from the moving components would be too great.

If the maximum BMEP is exceeded, the pressures inside the combustion chamber would be too great. This causes damaging loads on the head, piston, liner, and compression rings.

A turbocharger has a surge limit because the turbocharger is a dynamic machine that depends upon its output air as part of its powering mechanism. At low engine power, the turbocharger cannot pump air into the intake manifold above the limit line because the intake manifold pressure is too high. The turbocharger's air compressor will have no discharge flow and the part of the turbocharger's turbine will have no load and will speed up. Since the engine is not receiving enough air, its output horsepower has to drop below the surge limit. The compressor can start pumping again and put a load on the turbine. This will cause the turbocharger to slow down. This will happen again and again if you run the engine over the turbocharger surge limit. The turbocharger will increase and decrease rpm when you try to run the engine above this turbocharger surge limit. This varying rpm is known as surging. The variance in rpm can become quite large and set up damaging inertia loads in the turbine rotor and compressor impeller.

The same thing occurs in a centrifugal pump. The pressure that can be developed for every mass flow rate is limited.

Finally, at the upper-right-hand corner, where the BMEP and the rpm are at maximum, we are burning the maximum amount of fuel and releasing the maximum amount of heat. The temperatures and pressures in this region are so high that the engine materials and lubricating oil are destroyed. This is the thermal load limit for the engine.

The highest operating point of the engine is in the upper-right-hand corner of the diagram. It is selected to get an engine that is not too heavy. You will notice that we are not in the area of lowest BSFC. Owners could opt for a maximum operating point a little lower in the minimum BSFC area using a lower normal continuous rating. This means they would have a heavier, but more fuel efficient engine.

To calculate the parameters at lower loads, we use the propeller law. This law applies to fixed pitch propellers. It is based on the observation that the horsepower to drive a ship-shaped object through the water varies as the cube of the rpm of the propeller.

$$\frac{HP}{N^3} = Const$$

also $HP = \frac{PLAN}{33000}$ or for the same engine $HP = PConst_2 N$ $\quad Const_2 = \frac{LA}{33000}$

thus, $\frac{PConst_2 N}{N^3} = Const$ or combining constants $\frac{P}{N^2} = Const$

Since the performance diagram plots BMEP verses rpm, the fixed pitch propeller law has been modified to state BMEP varies as the square of the rpm.

As an example of the fixed pitch propeller law, an engine is running at 20,000 horsepower, 140 psi BMEP, and 100 rpm. What is the horsepower and BMEP at 50 rpm? The horsepower constant would be 1/50 and the BMEP constant would be .02 using the above units. The horsepower at 50 rpm is 2,500 and the BMEP is 35 psi.

Looking at the diagram you can see that the fixed pitch propeller law does not always go through the areas of lowest BSFC. That is one of the disadvantages of a fixed pitch propeller.

The controllable, reversible pitch propeller (CRPP) can overcome this problem, because the pitch controller can be programmed to load the engine so that the loading can follow the areas of lowest BSFC. Another advantage of the controllable reversible pitch propeller is that you do not have to reverse the engine to go astern. The controller just has to reverse the pitch while the engine is rotating in the original direction. However, this type of propeller costs more money. It is also less hydrodynamically efficient and requires more maintenance then a fixed pitch propeller.

Fig. 4.2 helps us choose a maximum or normal operating point for an engine and allows us to see what happens at lower loads. The manufacturer supplies other diagrams that help the operating engineer evaluate the engine to make sure it is running within prescribed parameters.

The curve shown in fig. 4.3 is for a constant speed application, such as a ship's service generator.

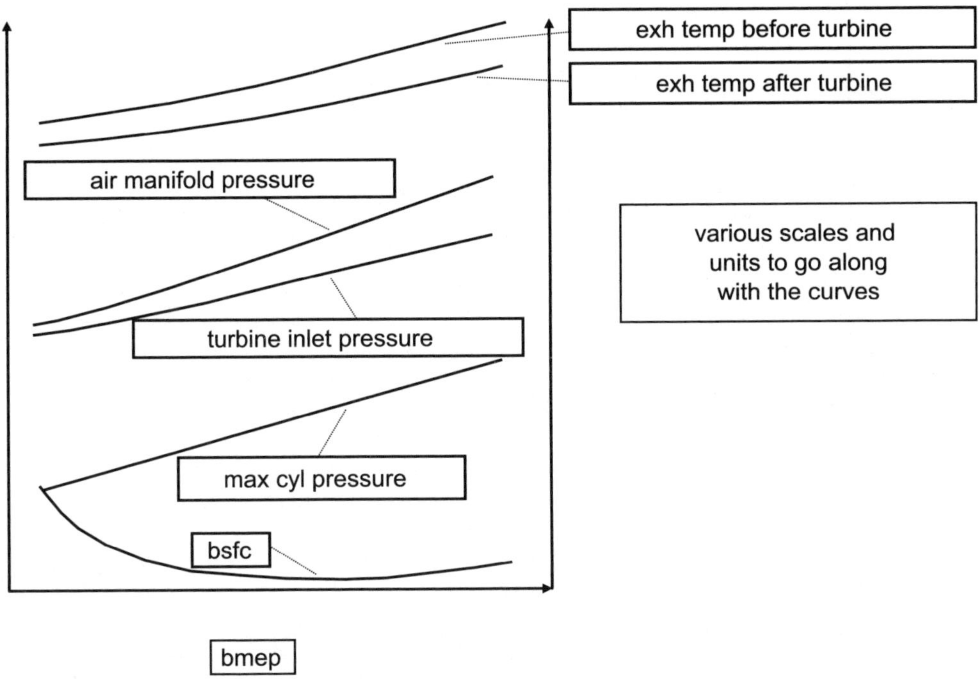

Fig. 4.3. Constant-speed performance curves

Trends to note in fig. 4.3:

1. The BSFC dips to its lowest point before the maximum load. This correlates with the previous performance curves.
2. The inlet temperature to the turbocharger turbine is higher than the outlet temperature, because energy is taken from the exhaust gas to run the turbine.
3. The turbine inlet pressure is lower than the air manifold pressure. This is true since the change in enthalpy is the parameter driving the turbocharger system.

We need a different set of curves to evaluate the performance of a main propulsion engine, since it does not run at a constant rpm. A way to evaluate this type of engine is to load it up, using the fixed pitch propeller law and varying the rpm.

As you can see in fig. 4.4, the propeller law loading accentuates the slope of the curves from low to high rpm which is also from low to high loads. This greatly narrows the area of lowest BSFC. You also can see that the turbocharger needs a high rpm to supply all the air at high-engine loads as opposed to low-engine loads.

Fuel Rate Calculations

We have been talking about the engine from the point of view of the engine manufacturer, because we have been discussing BMEP or brake horsepower. This horsepower is measured at the engine flywheel. This is because the engine manufacturers can measure the

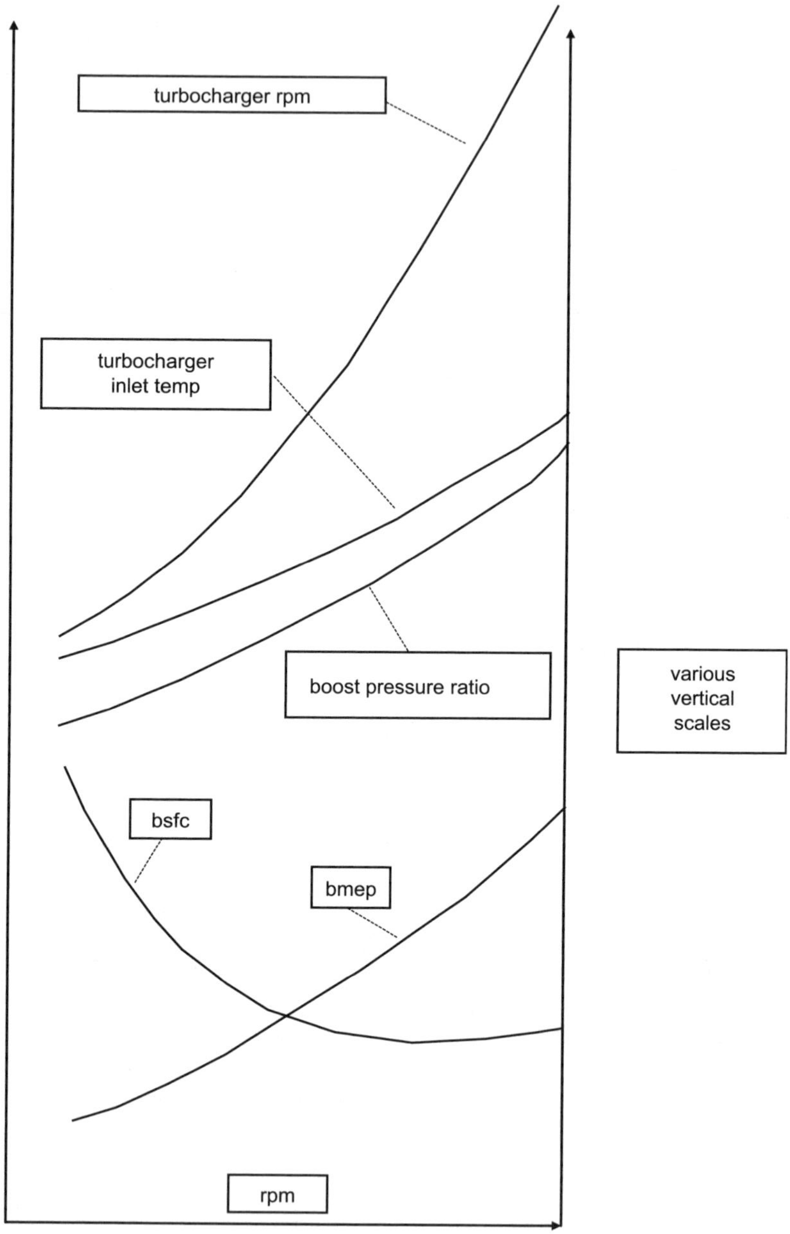

Fig. 4.4. Propeller law performance

horsepower at this point on their testbeds. They have no idea what you will hook the engine up to. In this example, it will be connected to a propeller shaft.

Naval architects calculate how much horsepower is required to drive the ship at a certain speed with the propeller turning at a certain rpm. To do this, they must take into account the hull resistance and the propeller hull interaction. Since the naval architect is calculating the horsepower required to drive the propeller shaft, this horsepower is called shaft horsepower. The architect's job is to find the power required to drive the propeller. The marine engineer's job is to calculate the brake horsepower at the engine flywheel from the shaft horsepower and pick an engine to satisfy this requirement.

Fig. 4.5 shows the various horsepowers in a ship.

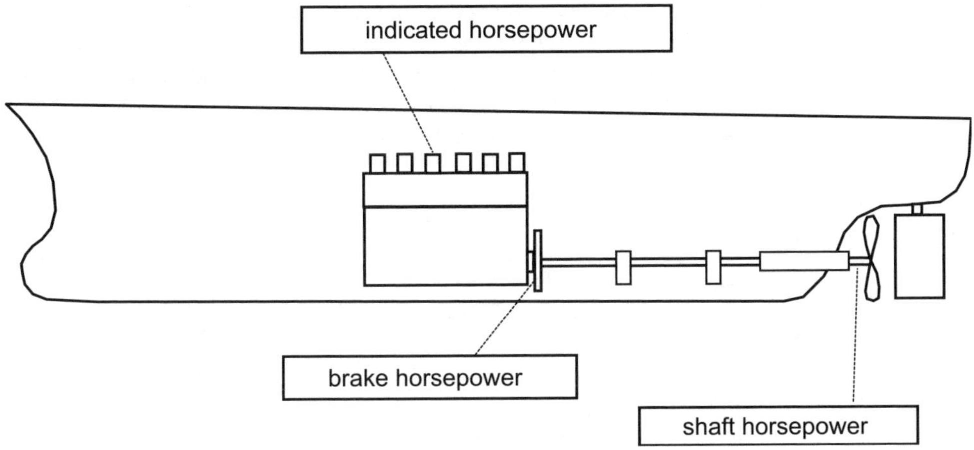

Fig. 4.5. Horsepower location

The shaft horsepower is set by the naval architect. The shaft bearings and stern tube have losses that must be added to the shaft horsepower to obtain the brake horsepower. Fig. 4.5 represents a slow-speed engine, which does not have a reduction gear. If this were a medium- or high-speed application, the reduction gear losses also would have to be added, and the power to drive any attached pumps. Other losses may have to be added, as seen below. The indicated horsepower is the horsepower delivered to the top of the pistons by the combustion gases. This is the highest horsepower of the system because the engine has internal friction losses.

In summary:

BHP= SHP + Reduction Gear Losses + Thrust Bearing Losses + Shaft Losses (Shaft Bearing and Stern Tube) + Possible Attached Pump Power + Possible Electric Drive Losses + Possible Shaft Generator Power

IHP= BHP + Engine Mechanical Losses

$$\text{Mechanical Efficiency} = \frac{BHP}{IHP} \times 100\% = \eta_{mech}$$

Marine engineers calculate the brake horsepower from the shaft horsepower for various engine configurations. They then go to various engine manufacturers with the brake horsepower and rpm. After choosing several engines that fit these basic criteria, they also consider fuel consumption and reliability before making a final decision.

This gives us a rough idea of what engine might be best for our particular application. To make a final decision, we must also consider what type of ship we are dealing with. The Society of Naval Architects and Marine Engineers (SNAME) Publication 3-27 outlines how to find the ship's fuel rate based on many particular parameters for a given ship. (SNAME 3-49 is the latest version.) In this way, we can evaluate our particular engine choices in a more detailed manner to pick the most economical engine for our ship.

Since one of the most important parameters we have to evaluate is the BSFC, we will take a closer look. The BSFC depends on the type of fuel the engine burned during the testing. This is because the heating value of the fuel can vary between 19,800 Btu/lb for No. 2 diesel fuel to 17,500 Btu/lb for Bunker C. The BSFC is the pounds of fuel burned per hour to produce one horsepower. However, if we are looking at Btu/in^3, then Bunker C would have the higher number.

The efficiency of the engine is independent of the fuel being burned, thus, we can write the following equation:

$$\eta = \frac{\text{output}}{\text{energy\_input}} = \frac{HP}{\dot{m}_f} \frac{}{LHV} \quad \text{Units} \quad \frac{HP}{\frac{lb_f}{hr}} \frac{}{\frac{BTU}{lb_f}}$$

However, this is not unitless, as efficiency has to be. Using the above parameters and units, we must insert the following conversion factor:

$$\eta = \frac{HP}{\dot{m}_f} \frac{}{LHV} \frac{2545 \quad BTU}{hr \quad HP}$$

Using the idea of BSFC we can write:

$$\dot{m}_f = bsfc \quad HP \qquad \text{Thus we have} \qquad \eta = \frac{HP}{bsfc \quad HP} \frac{}{LHV} 2545$$

For the same engine running with two different fuels at the same efficiency and horsepower, we have:

$$\frac{HP}{bsfc_1 \quad HP} \frac{}{LHV_1} 2545 = \frac{HP}{bsfc_2 \quad HP} \frac{}{LHV_2} 2545$$

Canceling out the like terms, we are left with:

$$bsfc_1 LHV_1 = bsfc_2 LHV_2 \quad \text{also} \quad \eta = \frac{2545}{bsfc}\frac{100\%}{LHV} \quad \text{with } \eta \text{ in } \%$$

As you see, the higher the heating value, the lower the BSFC. If engine manufacturers wish to look good, they always test their engines on No. 2 diesel fuel to get the lowest BSFC. If the engine is run on another type of fuel, we can always use the above equation to find the BSFC for that fuel.

Four-stroke Cycle Engine and Valve Timing

Now that we have worked with engines in a general sense and have the ability to choose an engine for a ship, we will study the different cycles in more detail.

Fig. 4.6 represents the pressure volume diagram of a four-stroke cycle engine. The work areas are represented by W1 and W2. W1 represents the work output of the engine. This is because it is enclosed by the power and compression strokes. W2 represents the work the engine has to perform to push the exhaust out and take the fresh air in. To find the net work out of the engine, subtract W2 from W1.

Net Work Out = W1-W2

This diagram represents a naturally aspirated four-stroke cycle engine. This is because the atmosphere line is in the middle of the intake and exhaust stroke. If this were a turbo-charged engine, two things would be different about the above diagram:

1. The whole trace would be shifted above the atmospheric pressure line, because the compressor side of the turbocharger pumps in air above atmospheric pressure. Large main propulsion engines commonly have intake manifold pressures above two atmospheres.

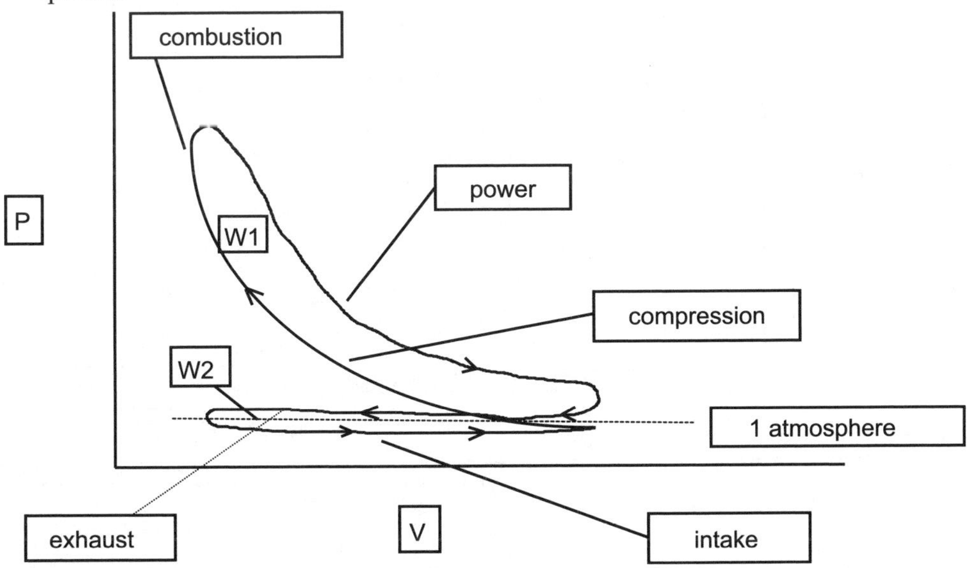

Fig. 4.6. Four-stroke cycle engine naturally aspirated

2. The intake trace would be above the exhaust trace. This is because the turbocharger is pumping in fresh air on the intake stroke that is higher than the exhaust stroke pressure. This means that during the intake stroke, the turbocharger actually is pushing down on the piston and putting more work into the engine than the other cylinders have to put into the engine during the exhaust stroke.

   Since the turbocharger in a turbocharged diesel engine is putting work into the engine during the intake and exhaust stroke, we have to modify the Net Work Out formula.

   $$\text{Net Work Out} = W1+W2$$

   For the engine to work, we must have valves to control the gas flow.

   Another way to represent the valve opening and closing is through the use of a polar timing diagram. This diagram is more accurate because the angles can be represented easily with this circular shape.

   As seen in fig. 4.8, events are not 180 degrees but the strokes are 180 degrees.

The open and closing of the valves do not take place at TDC or BDC for the following reasons:

1. The intake and exhaust valves are open during the overlap period to allow fresh air to purge out the clearance volume. Exhaust gas leaving the cylinder pulls in fresh air because of the exiting velocity of the exhaust gas at the end of the exhaust stroke in a naturally aspirated engine.
2. The exhaust valve opens before BDC of the power stroke for blowdown. At this point, the power stroke has little movement left before BDC, thus the exhaust valve opens, allowing exhaust gas to escape under its own pressure.
3. The intake valve closes after BDC because the air is still rushing in under its own inertia. The valve is timed to close just before the fresh air starts backing out of the cylinder, as the piston starts upward on the compression stroke.

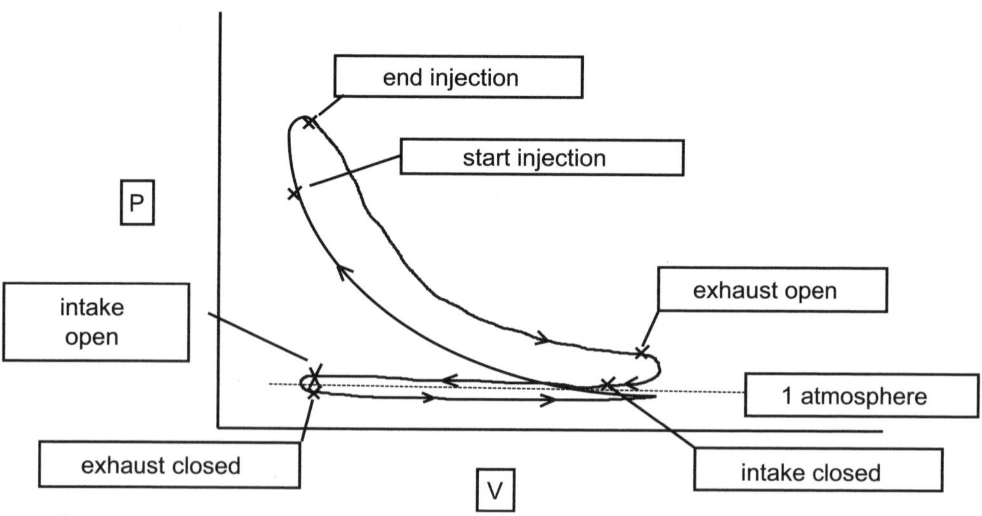

Fig. 4.7. Naturally aspirated four-stroke cycle pv diagram with valves

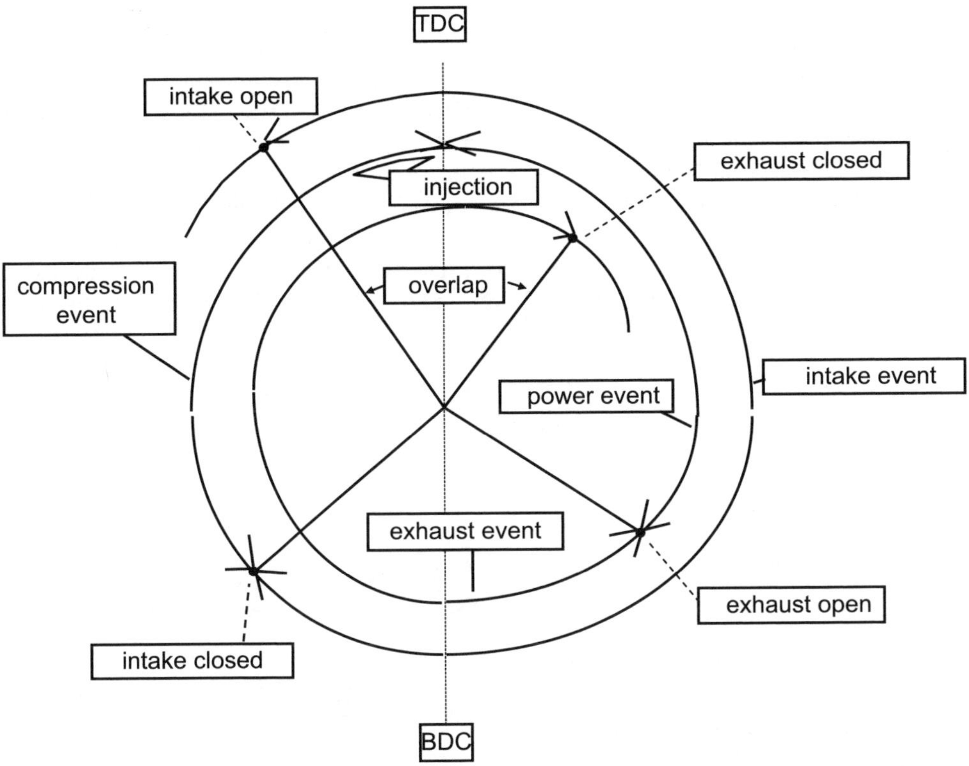

Fig. 4.8. Four-stroke cycle polar timing diagram

4. Fuel injection takes place just before TDC on the compression stroke. This is to allow time for the fuel to auto ignite and start to burn.

The angles of the valve opening and closing around BDC and TDC depend on the speed of the engine and whether it is turbocharged. Fig. 4.8 could represent either a turbocharged engine or a naturally aspirated high-speed engine.

1. A naturally aspirated four-stroke, high-speed diesel engine would have large angles around TDC and BDC. To have the same amount of time to exchange the gases at high rpm, the angles around the two dead centers must be larger.
2. In a turbocharged engine, the overlap is large so that cool air can blow through the clearance volume to cool off the engine internals. This is because the turbocharger enables the engine to run at almost double the horsepower for a given displacement and needs more cooling. The exhaust opens at a large angle before BDC to allow more energy from the exhaust gas to run the turbocharger. The intake valve will close at a large angle after BDC because the air is being pumped into the cylinder by the compressor end of the turbocharger. The air takes more time to start backing out of the cylinder as the piston comes up on the compression stroke.

In contrast, a diagram for a medium-speed, naturally aspirated diesel engine is shown in fig. 4.9. Notice the smaller angles TDC and BDC required around.

In both diagrams, the angles around BDC are a little larger than the angles around TDC. This is due to the angularity effect of the connecting rod and the crankshaft. Since the piston travels over half the stroke from TDC to 90 degrees, it does not travel as much near BDC, so the angles near BDC are larger for better breathing.

As seen in fig. 4.10, the piston travels more than one half of the downward stroke at 90 degrees.

If the connecting rod is lowered straight down to the center of the crankshaft, the piston would travel one half of the stroke. The lower end of the connecting rod cannot go to the

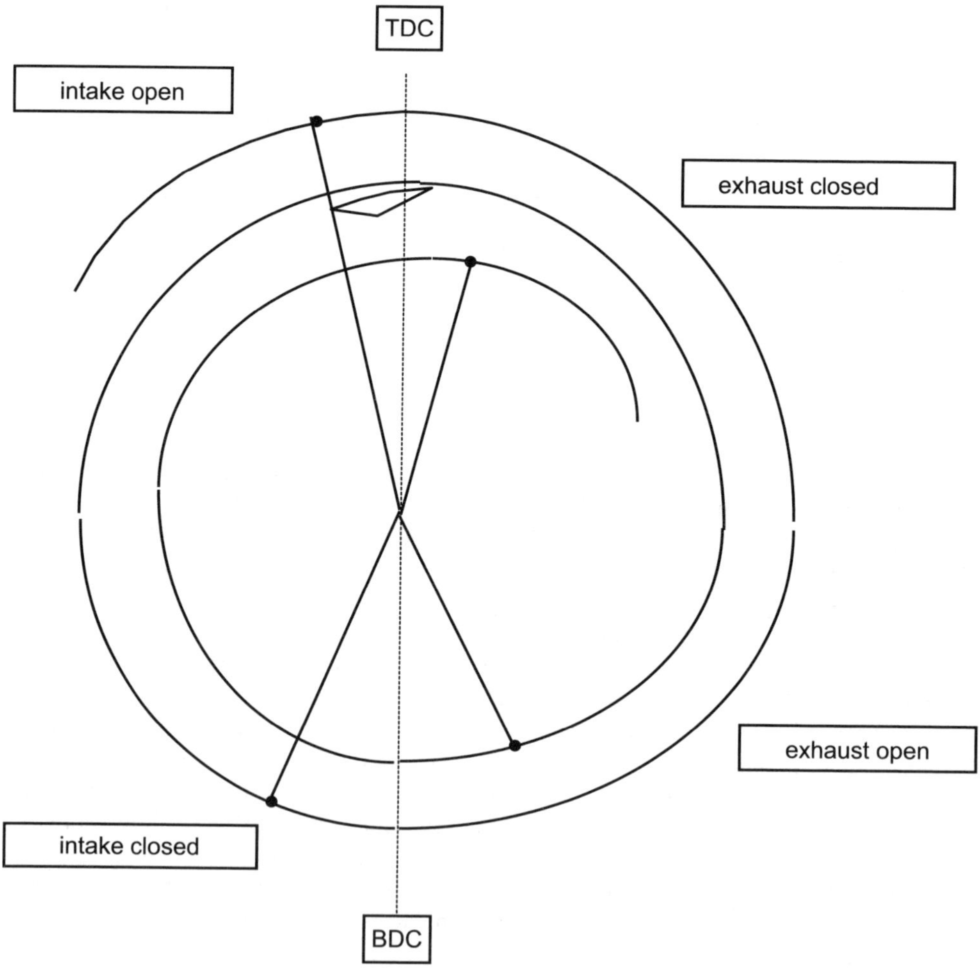

Fig. 4.9. Medium-speed four-stroke polar timing diagram

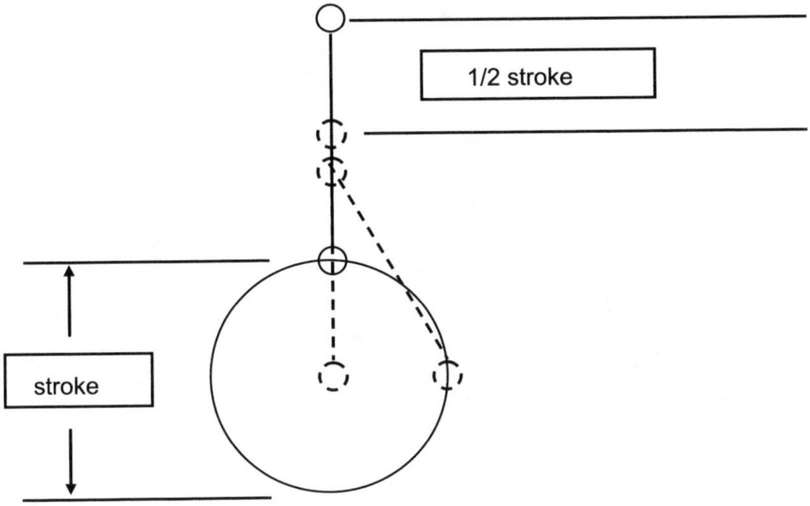

Fig. 4.10. Piston travel

center of the crankpin because it is attached to the crankpin. Thus, when the crankshaft is at 90 degrees, the lower end of the connecting rod must swing out and lower the piston more than one half of its stroke.

This means that the motion of the piston, as it travels up and down in the cylinder, is not sinusoidal. The motion is periodic. Fourier analysis can be used to emulate any periodic motion by the sum of sine waves at certain amplitudes. In our case, the general equation below does the job.

$$\text{Piston Position from TDC} = \underset{\text{Primary}}{A\text{cosine}(\theta)} + \underset{\text{Secondary}}{B\text{cosine}(2\theta)}$$

The piston position is made up of two terms. Both these terms are a function of the angle (θ theta) from TDC of crankshaft rotation. The first term has a frequency of crankshaft rotation. It also has the largest amplitude, which is why it is called the primary. The second term has a frequency of twice crankshaft rotation. It also has a smaller amplitude, and is called the secondary.

If we differentiate the equation twice, this is when we get primary and secondary acceleration. If we multiply the acceleration by the mass of the piston and the mass of part of the connecting rod, we get the force generated by the reciprocating components of the engine. This force can be separated into primary and secondary, depending upon which term in the original equation generated the acceleration. Primary forces have the frequency of the crankshaft rpm, while secondary forces have the frequency of twice the crankshaft rpm.

Crankshafts have general configurations so that the throws will balance out as many forces as possible. These forces occasionally rock the engine. The general crankshaft configurations will take care of this, if you have enough cylinders. Six is considered the optimal minimum number of cylinders. If you have less than six cylinders, you may have to

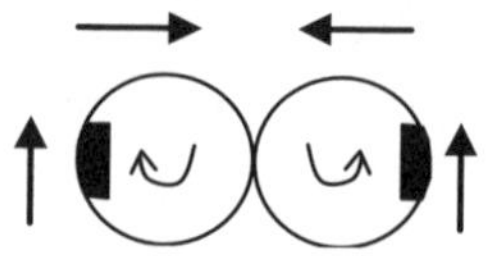

Fig. 4.11. Lancaster Balancer

add balance weights to the crankshaft for the primaries. No rotating member in the engine runs at twice crankshaft speed. Balance shafts running at twice crankshaft speed have to be added to some engines with less than six cylinders.

When you add a rotating weight, you add a force that has a horizontal and vertical component. The moving piston only has a vertical component. The weights added to the crankshaft have to be in opposing pairs. Balance shafts usually are mounted in counter rotating pairs. This is known as a Lancaster Balancer.

The unbalanced vertical force in the pair counteracts the piston and the horizontal forces in the pair balance each other.

Two-stroke Cycles and Scavenging Methods

We will now look at the two-stroke cycle port opening and closing. There are two types of two-stroke cycles engines. One is completely ported with intake and exhaust ports. The other uses intake ports and a timed exhaust valve in the cylinder head, run by the camshaft, just as the valves in a four-stroke engine operate.

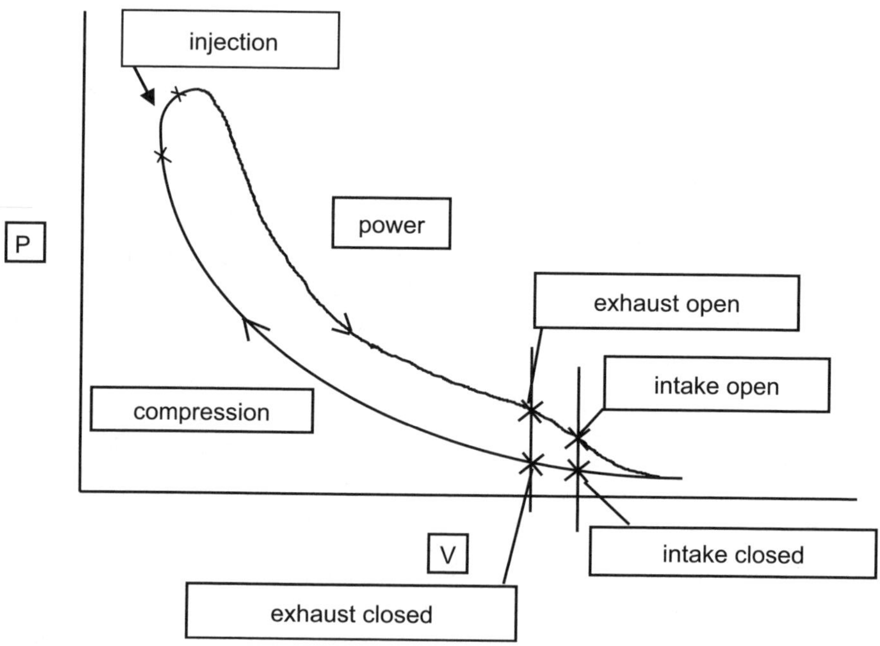

Fig. 4.12. Pressure volume diagram of completely ported two-stroke engine

The PV diagram in fig. 4.12 represents a completely ported engine. Notice that the ports open and close at the same volume point, because the ports are drilled into the cylinder liner wall.

The exhaust ports always open before the intake ports. This is to get the blowdown effect of the exhaust gas near BDC.

We can draw a polar timing diagram for this engine.

The problem with this type of engine is that the exhaust has to open before the intake for blowdown. This means the intake has to close before the exhaust, causing a problem with supercharging by the turbocharger. Air can escape through the still open exhaust ports as the piston rises.

This can be solved by using a timed exhaust valve, located in the cylinder head. The polar timing diagram of this engine is shown in fig. 4.14.

In the polar timing diagram shown in fig. 4.14, a two-stroke diesel engine with a timed exhaust valve, the exhaust valve opens first for blowdown and closes before the intake port closes. The turbocharger can pack more air into the cylinder for more effective supercharging by the turbocharger.

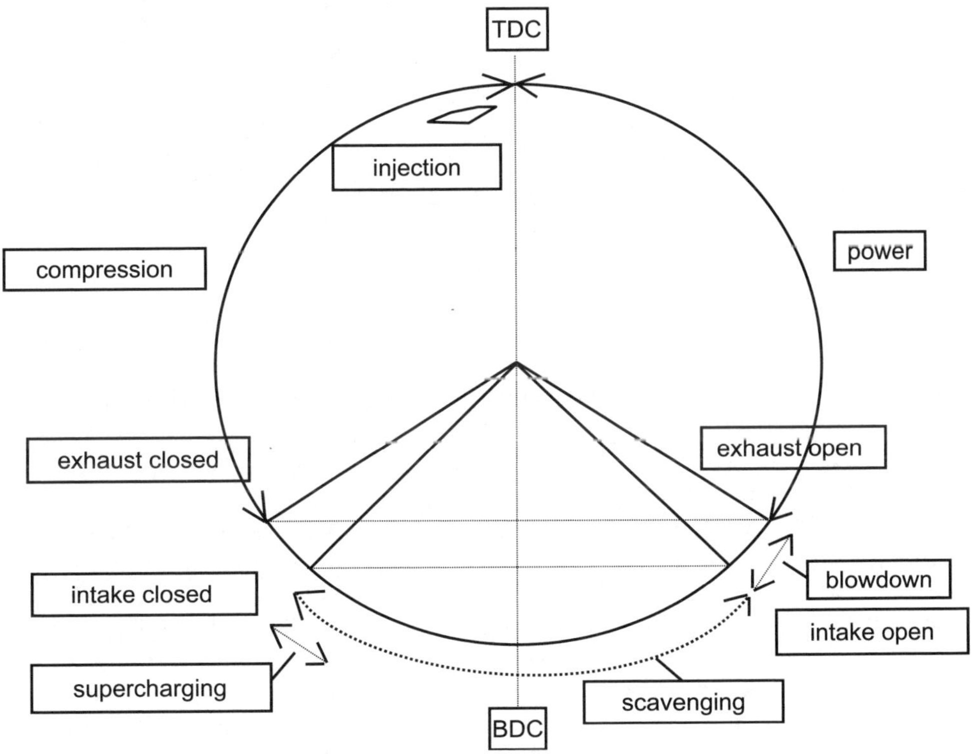

Fig. 4.13. Polar timing diagram of a two-stroke completely ported engine

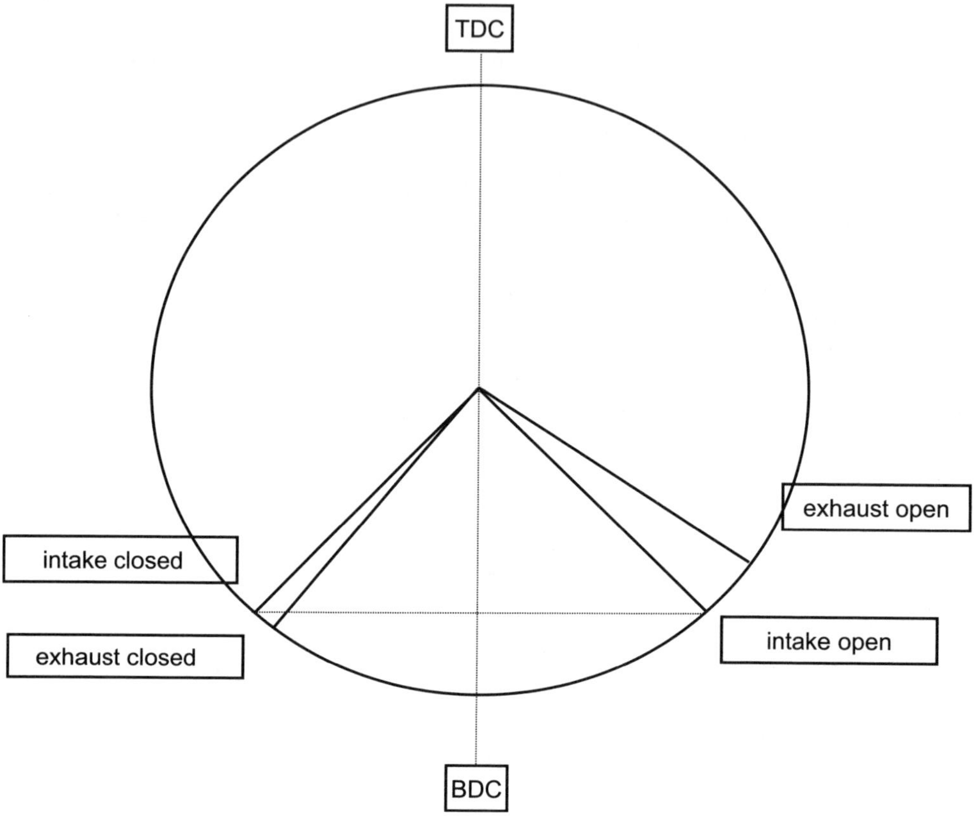

Fig. 4.14. Polar timing diagram of two-stroke diesel engine with timed exhaust valve

This type of scavenging is used on all modern, slow-speed diesel engines because it clears out the upper reaches of the combustion chamber most efficiently.

The scavenge methods in fig. 4.15 have their problems. The cross-flow method tends to leave combustion gases in the top of the combustion chamber. Part of the fresh air can blow right through. The loop-scavenge method is better, but it still tends to leave combustion gases in the top of the combustion chamber. The cross flow and loop are no longer manufactured because of this problem, but old Sulzer engines with the loop scavenge method are still in use. All new two-stroke cycle engines use the uniflow method of scavenging, since this purges most exhaust gases, even if you have a very long stroke with a very tall combustion chamber. This type of tall combustion chamber is used in all super-long-stroke diesel engines. This accounts for all of the new building in slow-speed engines because of the slow propeller speed and the overall thermal efficiency of the engine.

Another advantage of uniflow scavenging is the piston can be oil cooled. This is because only cool air flows over the piston. In a loop-scavenge engine, hot exhaust gases flow

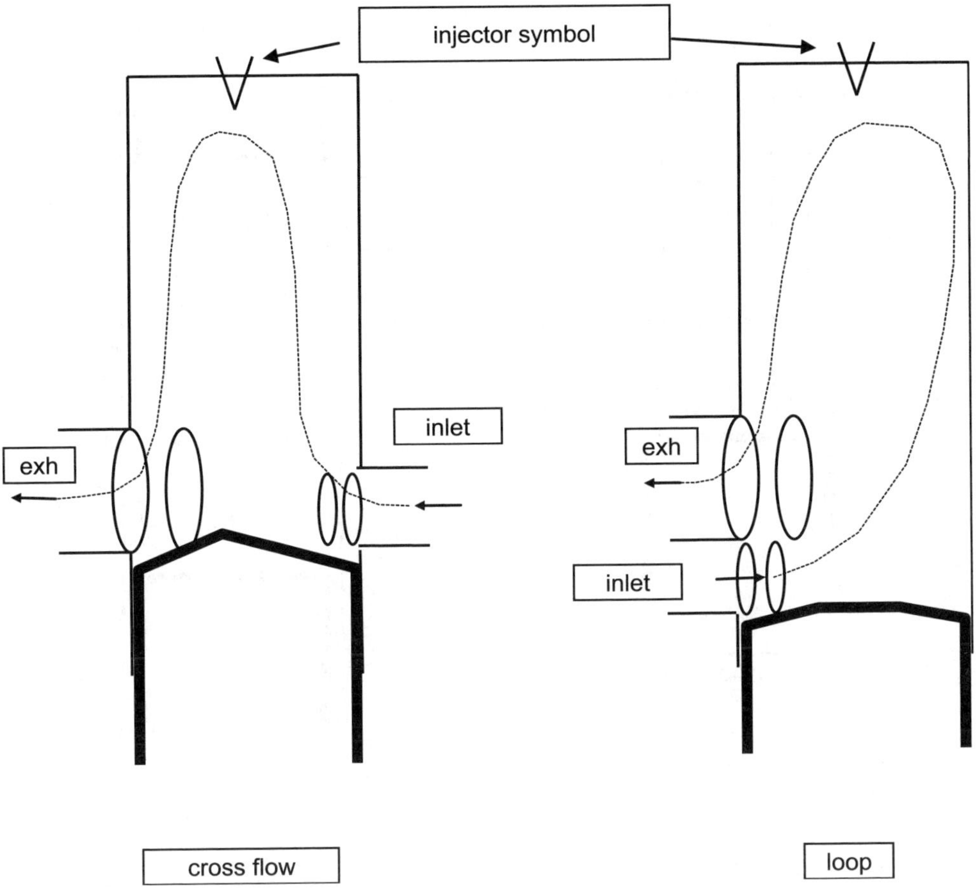

Fig. 4.15. Two-stroke cycle scavenging methods

over the piston during blowdown, so the piston must be water cooled. This necessitates the installation of a special piston cooling water system.

Turbocharger Summary

There are two types of turbocharger systems. One is pulse-type turbocharging, which is used on four-stroke cycle diesel engines. It uses the pulse of blowdown to run the turbocharger. The exhaust system is quite complicated since only cylinders firing at the same time can be led to the same nozzles in the turbine. If the exhaust from cylinders firing at different times are brought together, the exhaust could go from a cylinder with a higher exhaust pressure into a cylinder with a lower exhaust pressure. Several sets of exhaust pipe connect cylinders that have blowdown at the same time are connected to the same sets of nozzles in the turbine of the turbocharger.

The other type of turbocharger system is constant pressure. Here, all the exhaust is led to a large receiver where the pulses are turned into heat energy. The exhaust is then led to

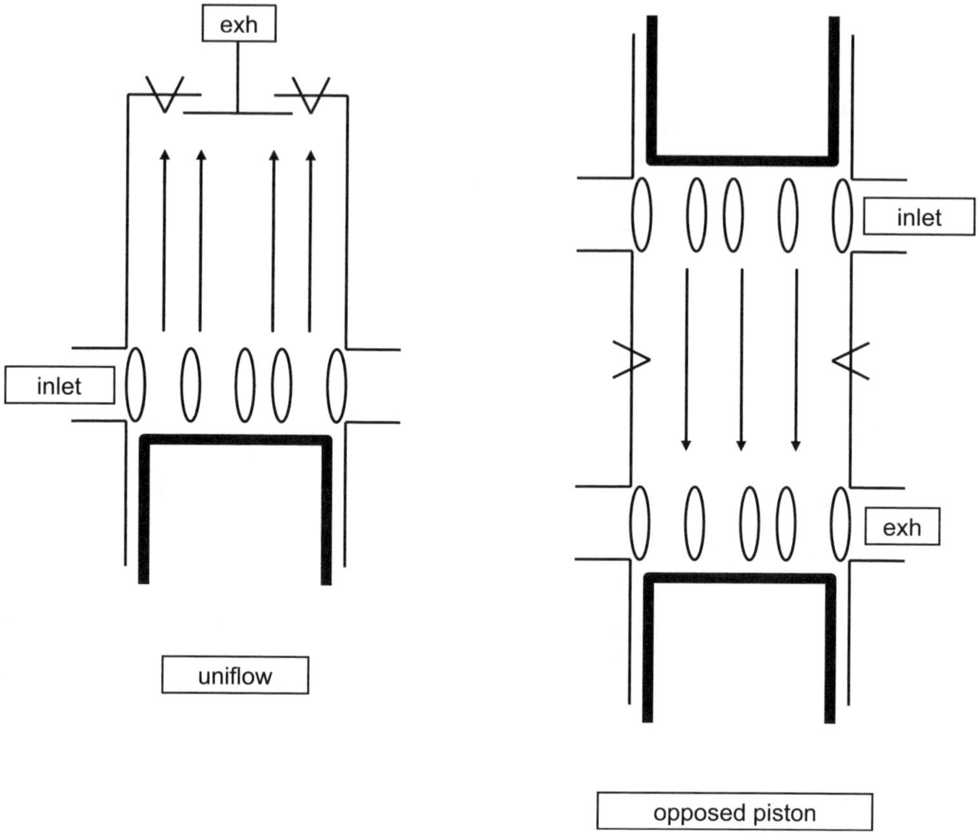

Fig. 4.16. Uniflow scavenge methods

all the turbine nozzles of the turbine of the turbocharger. This system is used on slow-speed engines. It is the most efficient type of turbocharging. This is because the velocity out of the turbine nozzles is constant and the blade angle of the turbine can be optimized for this velocity and the rpm of the turbocharger. In the pulse system, the blade angle can only be designed using an average velocity, since a pulse of gas produces a range of velocities.

Two-stroke cycle engines always need an air pump. Small engines usually use a roots blower for starting, and a turbocharger while under load. This makes for a more efficient and powerful engine. Slow-speed engines have an electric-powered fan and under the piston pumping for scavenge air during starting. When under enough load, the turbocharger can take care of all the scavenging needs and the fan is shut off. On some, engines doors open to disable the under piston pumping, and the air just circulates under the pistons. This makes for a more efficient engine.

Four-stroke cycle engines do not need an external air pump because the intake and exhaust strokes are an air pump.

The timed exhaust valve in all modern, slow-speed engines also aids in making the most of the turbocharger. The exhaust valve is tuned to open and supply energy to the turbine.

The exhaust valve is tuned to close before the intake port to allow the turbocharger to pack as much air into the combustion chamber as possible. The uniflow design also helps clear the exhaust gas from the combustion chamber. This adds to the initial cost and maintenance cost of an engine, but the costs are offset by the greater efficiency.

Rpm and Piston Speed

The higher the speed of an engine, the more horsepower and flow of reactants. However, the speed is limited by inertia forces, reliability, and economic life. The higher the speed, the more horsepower can be generated per given displacement, but the engine wears out faster.

The ratio of inlet air and exhaust passage area to cylinder bore is fixed, due to mechanical design considerations in the head. Eventually, any increase in rpm or piston speed results in a reduction of mean effective pressure, due to a reduction in trapped air density.

Mean piston speed can be calculated as follows:

mean piston speed = 2 x stroke x rpm

(ft/min) = 2ln (stroke in feet)

(ft/min) = ln/6 (stroke in inches)

Present Day Limits		
1000 rpm and higher	small high speed	2,000 to 2,400 ft/min
	medium speed	1,600 to 2,000 ft/min
300 rpm and lower	slow speed	1,200 ft/min

The above table shows that the slow-speed engine, which is the most durable, has the lowest rpm and piston speed. It gives the lowest velocity between the rubbing parts and the most thermally efficient. However, this engine cannot be put in a small truck. The other engines may be less durable and less thermally efficient, but they are the best engines in their applications, where weight is a big factor. A diesel engine, no matter what speed, is the most thermally efficient, internal combustion engine in its given application.

Study Questions

1. The brake horsepower divided by the indicated horsepower is the:
   A. thermal efficiency
   B. mechanical efficiency
   C. Carnot efficiency
   D. overall efficiency
2. Which engine is the most efficient?
   A. fast speed
   B. high speed
   C. medium speed
   D. slow speed
3. In diagram D, what number represents an area of lowest BSFC?
   A. 6
   B. 7
   C. 8
   D. 9
4. In diagram D, what number represents the turbocharger surge limit?
   A. 3
   B. 4
   C. 5
   D. 6

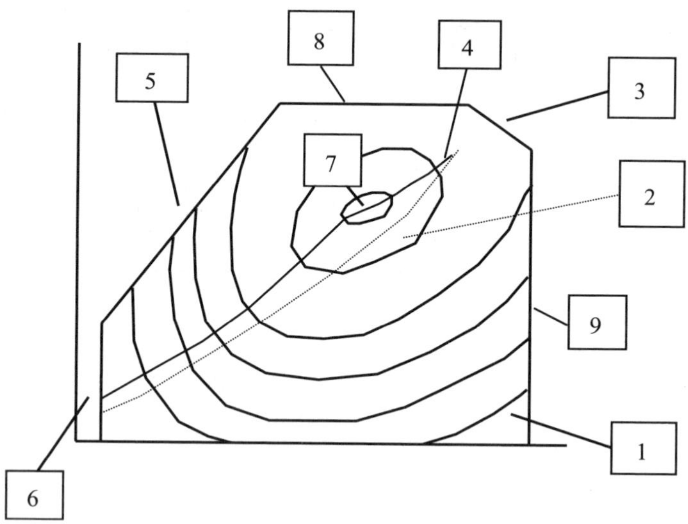

Diagram D

5. In diagram D, what number represents the thermal load limit?
   A. 3
   B. 5
   C. 6
   D. 8

6. An engine is running at 100 rpm and developing 25,000 horsepower when turning a fixed pitch propeller. What horsepower will the engine have to develop at 50 rpm?
   A. 2,000
   B. 2,500
   C. 3,125
   D. 12,500
7. Brake mean effective pressure is always:
   A. greater than the indicated MEP
   B. equal to the indicated MEP
   C. less than the indicated MEP
   D. the negative of the indicated MEP
8. Which of the following will increase the thermal efficiency of a diesel engine?
   A. decrease the compression ratio
   B. add an engine driven supercharger
   C. add a turbocharger
   D. change from distillate to bunker fuel
9. Which engine uses a crosshead?
   A. fast speed
   B. high speed
   C. medium speed
   D. slow speed
10. An engine manufacturer claims a BSFC of .3 (lb/HP hr) burning 19,800 Btu/lb fuel. The owner is going to run the engine on 17,500 Btu/lb fuel. What would be the BSFC?
   A. .265
   B. .34
   C. .355
   D. .36
11. What type of fuel is the owner going to burn in the engine?
   A. gasoline
   B. No. 2
   C. kerosene
   D. Bunker C
12. Diagram A represents what type of engine?
   A. one-stroke cycle
   B. two-stroke cycle
   C. four-stroke cycle
   D. five-stroke cycle

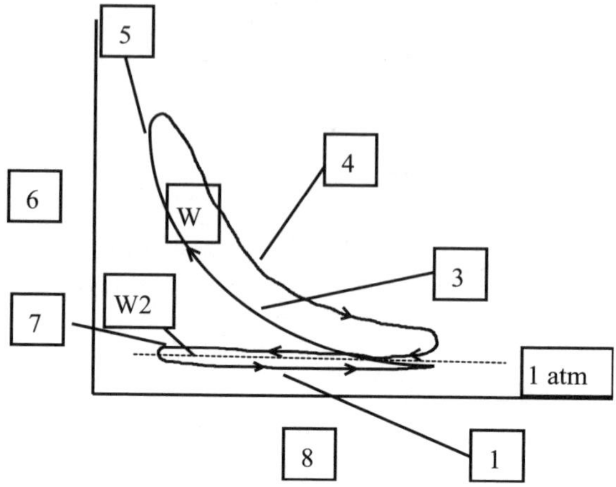

Diagram A

13. Which number in diagram A represents exhaust?
    A. 1
    B. 3
    C. 4
    D. 7
14. To obtain the net work out of the cycle in diagram A:
    A. add area W1 to W2
    B. subtract area W2 from W1
    C. subtract area W1 from W2
    D. multiply area W1 by W2
15. Between what two strokes does scavenging of the clearance volume in a four-stroke engine take place?
    A. intake and power
    B. exhaust and intake
    C. power and exhaust
    D. power and compression
16. Injection is represented by which number in diagram B?
    A. 2
    B. 3
    C. 4
    D. 5

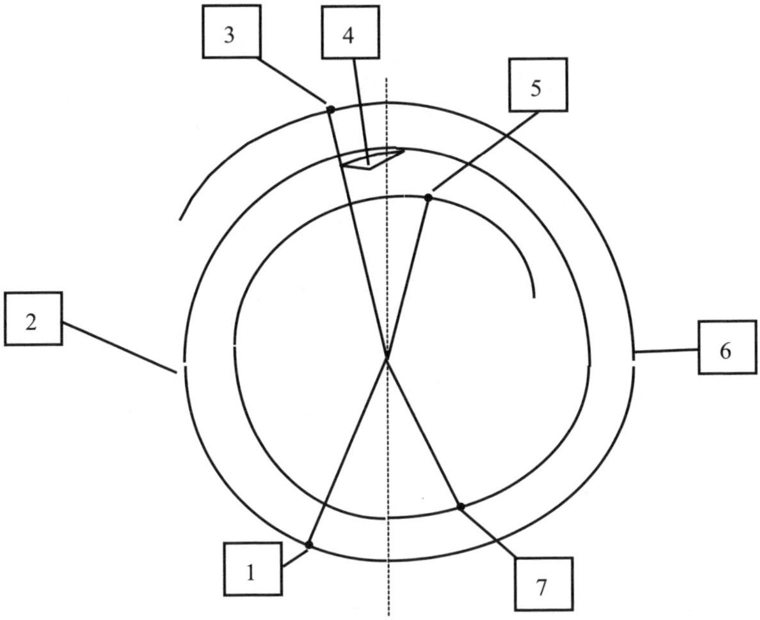

Diagram B

17. Diagram B represents which type of engine?
    A. turbocharged slow speed
    B. turbocharged medium speed
    C. non-turbocharged slow speed
    D. non-turbocharged medium speed
18. Overlap is represented between which numbers in diagram B?
    A. 1-2
    B. 3-4
    C. 3-5
    D. 4-5
19. At what number in diagram B is some extra air being packed into the combustion chamber?
    A. 1
    B. 2
    C. 3
    D. 4
20. Piston motion is:
    A. cyclic sinusoidal motion
    B. cyclic non-sinusoidal motion
    C. cyclic linear motion
    D. cyclic parabolic motion

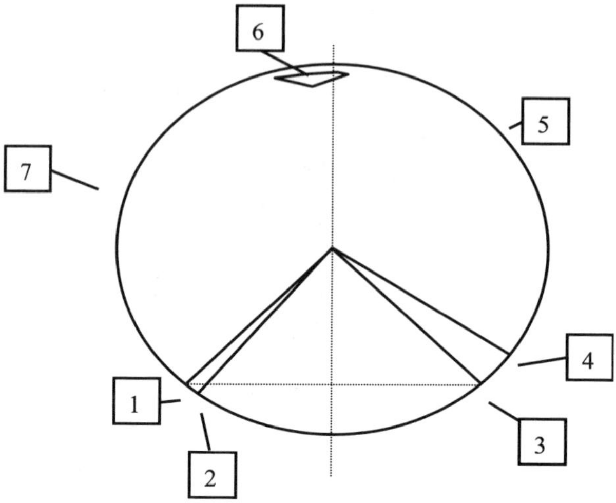

Diagram C

21. Which engine is represented by diagram C?
    A. completely ported
    B. loop scavenged
    C. uniflow scavenged
    D. cross-flow scavenged
22. In diagram C, scavenging takes place between __________:
    A. 1 and 2
    B. 2 and 3
    C. 3 and 4
    D. 4 and 5
23. Which diagram represents an engine that needs an air pump?
    A. B
    B. C
    C. B and C
    D. none of the above
24. Which type of engine must use water cooled pistons?
    A. crossflow
    B. loop
    C. uniflow
    D. two-stroke
25. Which type of turbocharger system is used on slow-speed engines?
    A. pulse
    B. constant pressure
    C. both
    D. neither

26. In diagram C, blowdown starts at:
   A. 1
   B. 2
   C. 3
   D. 4

27. Which diagram represents an engine with a intake valve?
   A. B
   B. C
   C. B and C
   D. none of the above

28. What type of scavenging is used on all new slow-speed engines?
   A. crossflow
   B. loop
   C. uniflow
   D. two-stroke

29. Which type of turbocharger system is used on four-stroke cycle engines?
   A. pulse
   B. constant pressure
   C. both
   D. neither

30. A four-stroke cycle engine is running at 2,000 rpm with a bore of 4 inches and a stroke of 6 inches. What is the mean piston speed in feet/minute?
   A. 2,000
   B. 2,400
   C. 4,000
   D. 4,800

Chapter 5
Thermal Loading

Thermal Loading of the Combustion Chamber

The upper-right-hand limit in the operating area of the engine performance curves is the thermal load limit. Near this limit, the engine burns the maximum amount of fuel and produces the maximum BMEP and power. Past this limit, the exhaust valves last only a few hundred hours, turbine nozzles and blades are short lived, and high heat flows occur in the head and piston. We want to keep temperatures in the combustion chamber as high as possible for the best thermal efficiency, but we do not want to destroy the engine. We will now look at some of the problems of running an engine at high temperature.

Creep. This is a term used to describe the loss of strength in a material at elevated temperatures and the deformation when a force is applied.

Thermal Stresses. These stresses are generated inside a material by uneven heating. The hotter material literally tries to expand away from the cooler material building an internal stress.

Stress Riser. When a sharp change in geometry from one form to another occurs, the lines of force are concentrated along the sharp edge. This means that the force is concentrated in a smaller area and the stress increases. As seen in fig. 5.1, the lines of force are evenly distributed in the object, both top and bottom. The stress is higher in the bottom because the area is smaller. However, in the sharp-cornered transition piece, the stresses are higher because the forces are concentrated in a smaller area. If fillets are added, then the forces smoothly flow from one area to the other.

If you have a crack on the surface of an object, a stress concentration also can occur.

Materials exposed to high temperatures oxidize. The products of oxidation are always weaker than the original material.

The lubricating oil film is destroyed at high temperatures. This promotes wear between the moving parts, such as the cylinder liner, the piston rings, and the piston.

In practice, the failure of a part usually starts with the surface material weakening because of heat, then a crack forms. This produces a stress riser. The thermal stresses produced during overheating causes the crack to propagate. The heating and cooling of parts during normal startup and shutdown propagate the crack, eventually leading to failure.

To predict the level of thermal stresses, the isotherms must be predicted and then backed up by measurements. From these isotherms, stresses can be calculated. These will be the thermal stresses as the hot material tries to expand away from the cold material. This used to be very hard to do. Calculus can be used to predict the isotherms and the thermal stresses.

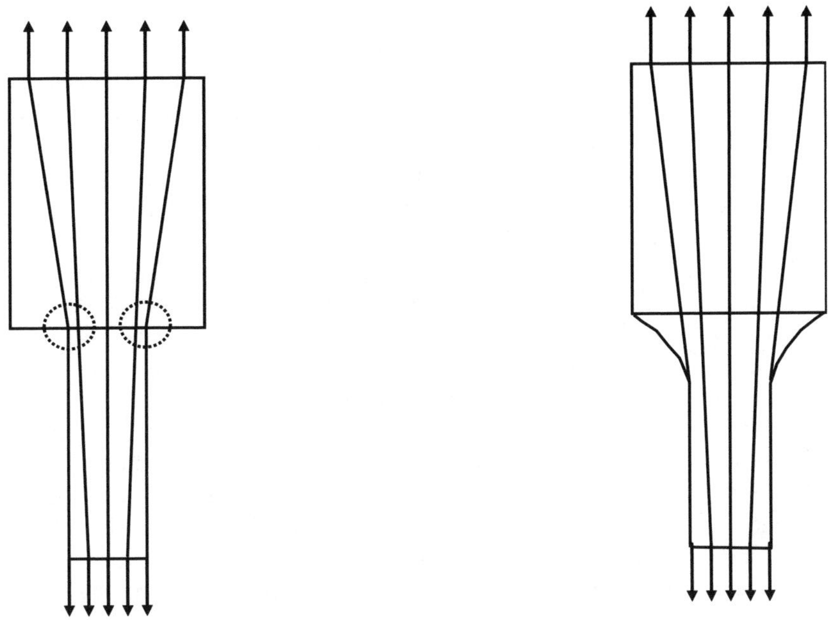

Fig. 5.1. Geometric stress riser

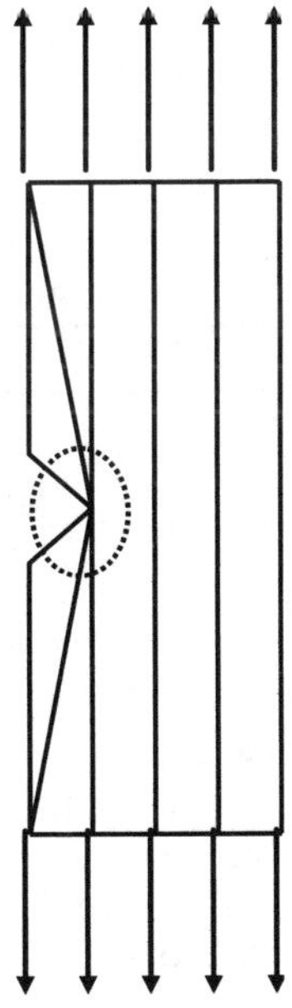

Fig. 5.2. Crack stress riser

Now, using finite element theory, the isotherms and, the thermal stresses can be calculated to a fair degree of accuracy. It takes the solution of thousands of simultaneous equations to accomplish this feat. We also must add in the stresses produced by the mechanical load of the combustion gas pressure. Naturally, we can do this calculation only with computers. Even then, it may take the computer hours or days to solve all the simultaneous equations. In the past, you were faced with evaluating thousands of points of temperature and thermal stress. Now, the computer can plot a colored representation of the part showing the hottest temperatures and the highest thermal stresses.

These finite element programs are getting better because more sophisticated measuring techniques are being used to check the theoretical values and fine-tune the program. The use of thermocouples and remote telemetry can measure the isotherms on moving parts. Strain gauges and remote telemetry can also cross-check the actual strain produced by the thermal stresses and mechanical loads of the combustion gases.

A general trend can be seen in the geometry of engine parts that are subject to thermal stresses. This trend is due to the fact that the thicker the part, the higher the thermal stresses can be. However, to make a part strong, you want to make it thicker. The only way you can get around this is to make the part from a stronger material. This usually means a more expensive material and manufacturing technique.

For example, cast iron is relatively weak, but with casting you can make a complex part quickly and cheaply. Cast iron is like particle board (glue and sawdust). If you had a 1" x 1" made from particle board, you could easily break it. If you had a 1" x 1" piece of wood with the grain intact, you would have a hard time breaking it, especially if it was oak. This continuous grain wood is like a forging. Steel is pounded and then machined into final shape, creating a very strong part costing more money. Because we have a strong material, we can make the part thinner and cut down on thermal stresses.

In general, the parts of an engine around the combustion chamber are made of the following materials:

cylinder head	
cast iron	lower output engine
nodular cast iron	
alloy cast iron	
alloy cast steel	higher output engine

Lower output engines use cast iron heads. This is a cheap way to make a very complicated part with many passages. The higher the output of the engine, the more exotic the material, as seen with the alloy cast steel.

The slow-speed engine has such a large head that the combustion gases produce extremely large forces. This pressure working on a large area produces tremendous stresses inside the material of the head, in addition to the thermal stresses. To keep the head from failing, it usually is made of forged alloy steel, which is bore cooled so that the thermal stresses do not cause the head to fail. Bore cooling is the drilling of many small water passages in the head near the combustion chamber. This allows for a thin thermal thickness, backed up by a thick layer of forged steel not subject to a large temperature gradient.

Many such sets of small passages are drilled into the forged steel head of a slow-speed engine to affect bore cooling.

The area between exhaust valves is another problem in trunk piston engines. This is an area of high heat flow and many heads fail with a crack starting from an exhaust valve opening. This can be solved with sufficient cooling in the area around the opening and rounding off the openings, as shown in fig. 5.4.

The flame plate is the area in the head right above the combustion chamber. The water circulation in the head above this area must be well designed to carry away any undissolved gases in the cooling water from this area. These undissolved gases insulate the flame plate and cause it to fail. Most heads also have a vent leading to the head tank of the cooling water system to lead these gases away.

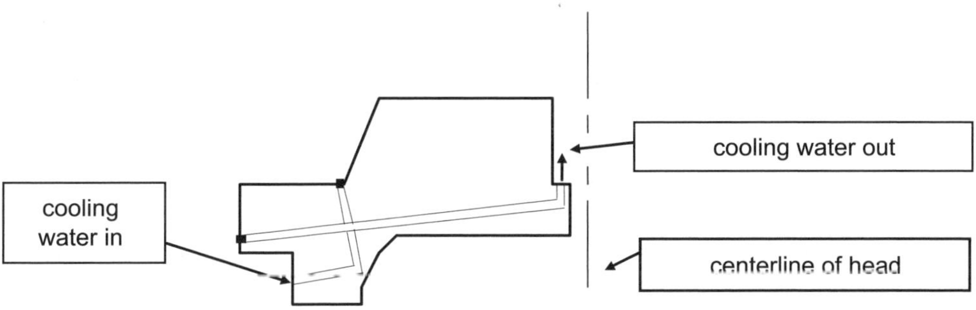

Fig. 5.3. Forged slow-speed engine head with bore cooling

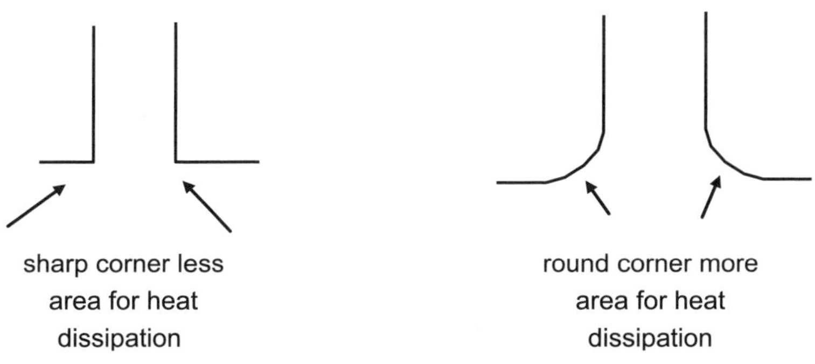

Fig. 5.4. Valve opening configuration

Piston Construction

Cast iron is used in low output, slower engines. Cast iron with steel crowns is used in higher output, slower engines.

Aluminum alloy is used in high speed and medium speed engines of today. Aluminum alloy with steel crowns is used in high-output, high-speed, and medium-speed engines of today.

In piston material, it is desired to have good thermal conductivity for heat dissipation, high strength, and light weight. The light weight is needed in high- and medium-speed engines because of the reciprocating motion of the piston. A lighter material produces less inertia forces on the piston and connecting rod. The thermal conductivity of aluminum is high, helping it to dissipate heat easily. In the ring groove area, aluminum has less pounding resistance then cast iron.

A problem with aluminum is its high coefficient of expansion. This is managed by having larger cold clearances between the piston and the cylinder liner than you would have with cast iron pistons. In large aluminum pistons, the piston ring grooves are contained in a Ni-resist cast iron insert. This solves the pounding resistance problem of aluminum. The Ni-resist cast iron has a higher coefficient of expansion compared to normal cast iron, thus, it matches the expansion of the aluminum better than the normal cast iron.

Piston Cooling

Ways to dissipate heat in a trunk piston:

- through the oil film to the cylinder wall to the jacket cooling water
- lubricating oil thrown up from the crankshaft to the underside of the piston (The crankshaft never dips into the oil in the sump, splashing it around. This would cause the lubricating oil to be aerated and reduce its lubricating qualities.)
- oil jet
- cocktail shaker
- oil cooling coils cast into the piston
- Ways to dissipate heat in a crosshead piston
- through the oil film to the cylinder wall to the cooling water
- oil cooled through telescopic tubes or through the piston rod for uniflow-scavenged engines
- water cooled pistons through telescopic tubes for loop scavenged engines

The crosshead engine uses a piston rod seal, which prevents any crankcase oil from being thrown up from the crankshaft.

Star Cracking in the Piston Crown

The temperature distribution in the piston crown has both a radial and axial gradient. If the piston is overheated or poorly designed, the following may happen:

1. The hot material expands with the cooler material around and below restraining it.
2. The hot material is then in a state of compression.
3. The hot material will creep-deform, due to losing its strength because of high temperature and the compression is relieved.

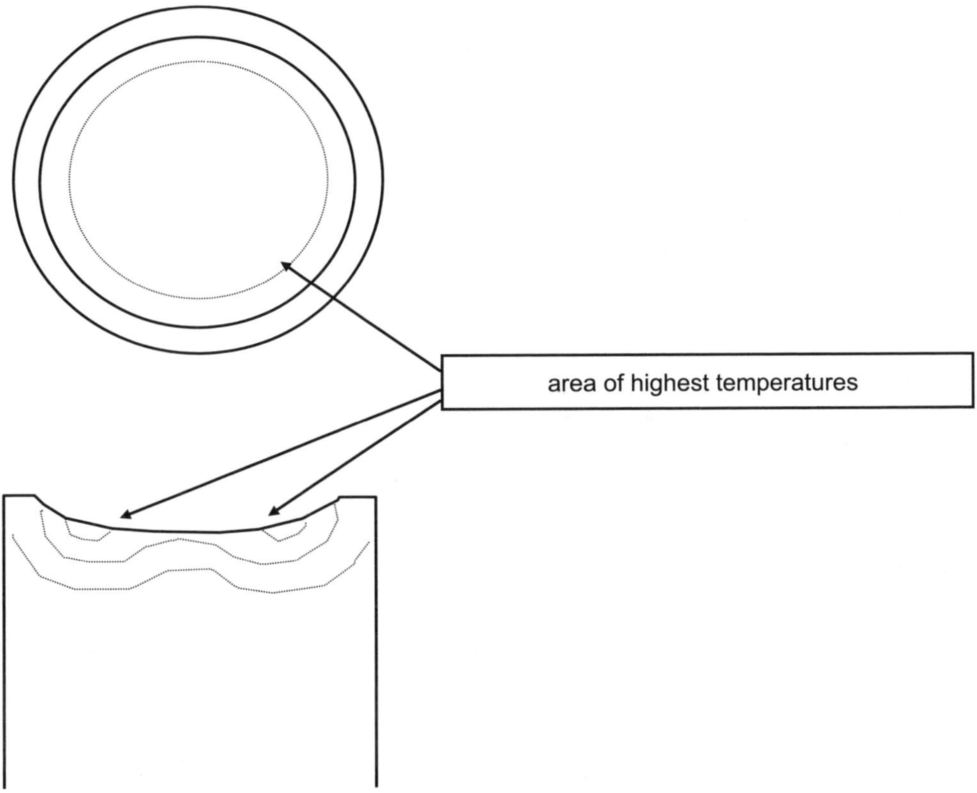

Fig. 5.5. Piston temperature distribution

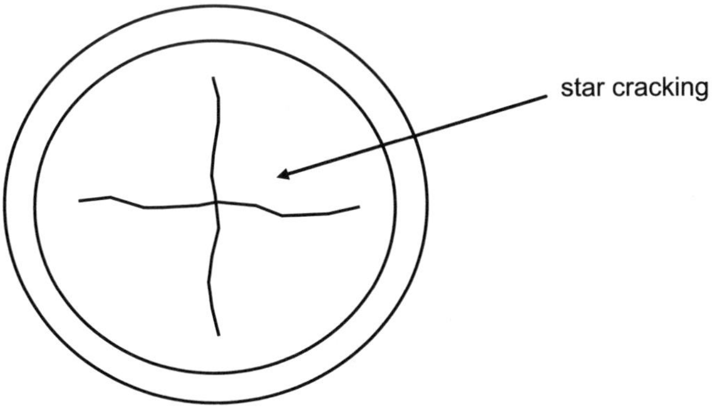

Fig. 5.6. Star cracking in pistons

4. When the engine shuts down and cools off, the formerly hot area goes from a state of compression to a state of tension.
5. A crack may initiate. With more cycles of start-up, overload, and shutdown, the crack may propagate, resulting in piston failure.

Piston Wall Stresses

Because of the temperature gradients shown in fig. 5.7, the piston deforms.

The gas pressure will produce the deformation shown in fig. 5.8.

With the cyclic loading of the gas pressure, the material under the rim of the crown may fatigue and start a crack, leading to piston failure.

Piston Ring Groove Cooling

The piston ring grooves should have a temperature less than 200 °F. This is to promote a good oil film and prevent the oil from oxidizing. To aid in this cooling, the top land is made thicker and a heat dam is used to retard heat flow to the ring grooves.

As you can see, the heat dam can weaken the piston crown so it must be carefully designed. Another way to have a heat dam and a strong crown is to use a cast in the cooling coil. This is shown in the second diagram.

In high-output engines, the crown is made of high-strength, heat-resistant alloy steel. This makes the top diagram even easier to design, without precipitating a failure. The rest of the piston can be made of a cheaper material, such as cast iron or aluminum.

Valves

The heat load on intake valves is minimal, so their design is not a problem. The exhaust valve is a problem, not because of the heat load on the valve material, but because of heating the fly ash. Vanadium and sodium salts combine to form a eutectic that melts at about 600 °C. An eutectic material is one that has a melting point below the melting point of its constituents. Water and ethylene glycol is an example. These form exhaust valve deposits

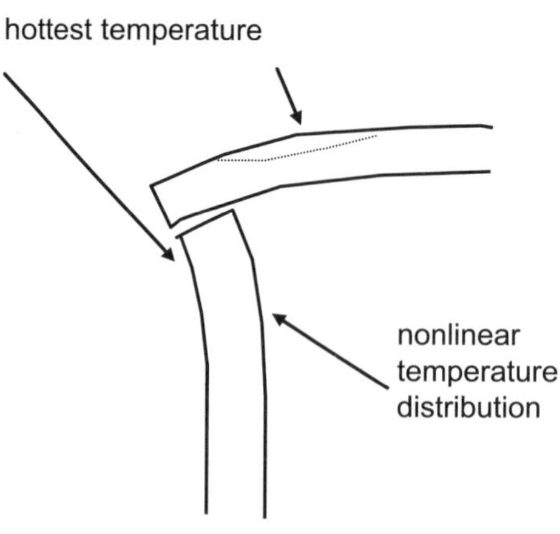

Fig. 5.7. Piston temperature gradient

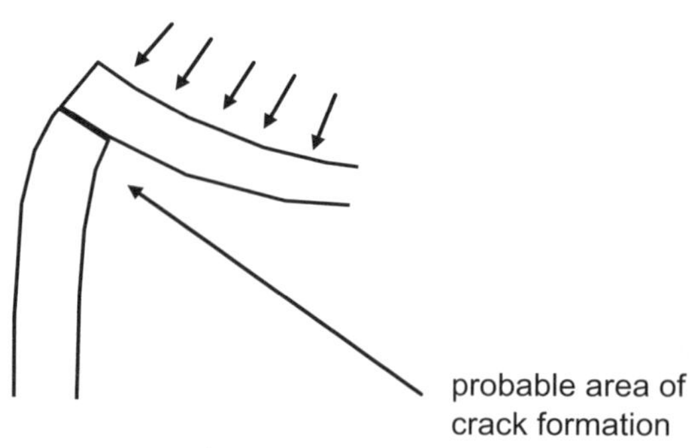

Fig. 5.8. Piston gas pressure deformation

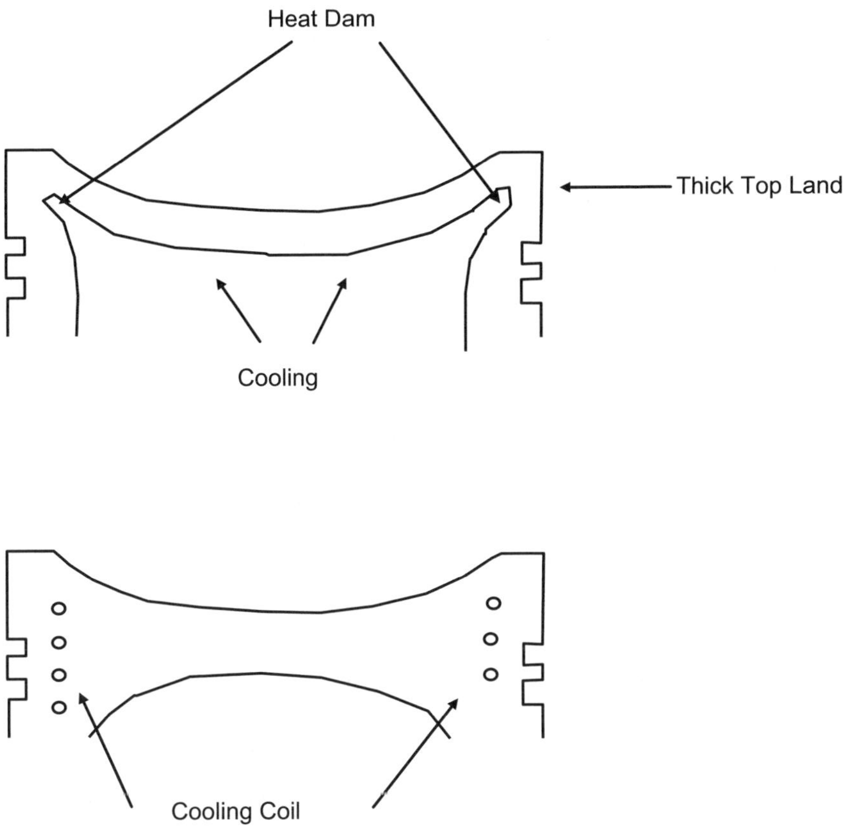

Fig. 5.9. Piston heat dams

that crack, leaving a gap that hot combustion gases rush through and erode the valve seat. This is called wire drawing. The exhaust valve seats must be cooled to around 550 °C, and the fly ash will bounce off and not stick to the valve seat. This problem occurs when using heavy fuel. Distillate fuel has very little, if any, fly ash content.

All valves are cooled through their seats when the valve is closed. The valve seats are also cooled through conduction up the stem to the valve guide. In turbocharged engines, both valves can be cooled by air blowing by them during the overlap period.

Exhaust valves need extra cooling because of the exhaust valve deposits. Listed below are ways to specifically cool exhaust valves:

1. The valve seat insert in the cylinder head can be in direct contact with cooling water. This insert is made of stellite, a temperature-resistant cobalt-based steel alloy.
2. Valve rotators can be used to rotate the valve and help knock off deposits.
3. Sodium-filled valves can use convection currents to move the heat into the valve guide area. Sodium is a liquid at the operating temperatures of the engine.
4. The valves can be water cooled with flexible hoses.

These are listed in order of reliability. The water cooled valve seats have no moving parts, while the flexible hoses may fatigue and leak. The moving parts of some valve rotators may fail. The sodium-filled valves may break and the sodium will react violently with any water in the combustion chamber.

Liners

The most efficient shape to contain pressure is the sphere. The second-most efficient shape is the cylinder. This means that all cylinder liners can be made of cast iron. Though the cast iron is weak, casting is a cheap manufacturing technique. The cast iron also makes a good bearing surface for the piston and piston rings. Cast iron has large grain boundaries that can absorb oil and help with surface lubrication. However, in slow-speed engines, the cast iron has to be very thick near the top of the combustion chamber. This is because of the high pressures of combustion acting on large areas. With these high pressures come high temperatures, thus, we have a problem with thermal stresses. As in the heads of slow-speed engines, bore cooling is used in the tops of the cylinder liners.

Fig. 5.10 shows only two sets of small water passages in the thickest part of the liner. Many sets of these passages are located around the top of the liner.

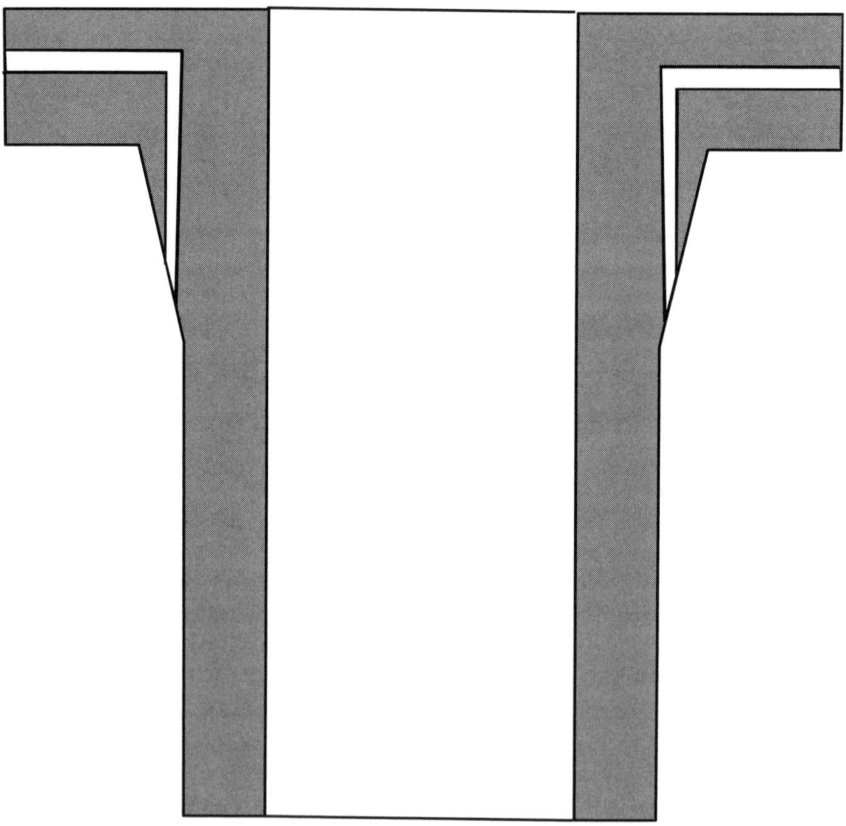

Fig. 5.10. Cylinder liner bore cooling

Study Questions

1. Creep is a term used to describe the loss of ________ in a material at elevated temperatures.
   A. strength
   B. heat
   C. connectivity
   D. conductivity
2. The unit or units of stress are:
   A. force
   B. force/area
   C. pressure/area
   D. force/volume
3. Which engine uses bore cooling of the cylinder cover?
   A. trunk piston
   B. crosshead
   C. high speed
   D. medium speed
4. Star cracking occurs in the:
   A. piston crown
   B. piston skirt
   C. cylinder head flame plate
   D. top of cylinder liner
5. Which valve is prone to deposits?
   A. air start
   B. fuel
   C. intake
   D. exhaust
6. The property of a material that directly affects thermal stresses is:
   A. specific heat
   B. coefficient of expansion
   C. specific gravity
   D. coefficient of heat transfer
7. What material is a crosshead engine cylinder head made from?
   A. cast iron
   B. nodular cast iron
   C. alloy cast steel
   D. forged steel
8. Which piston must have larger cold clearances?
   A. cast iron
   B. steel crown
   C. aluminum
   D. forged steel

9. What prevents the heat of combustion from overheating trunk piston compression rings?
   A. heat block
   B. heat dam
   C. bore cooling
   D. heat conduction
10. Which element helps form a eutectic in valve deposits?
   A. sodium
   B. chlorine
   C. calcium
   D. sulfur

Chapter 6

Intake and Exhaust Systems

System Design

A number of general steps are taken in designing a system. These steps can be followed in the design of any fluid system for a diesel engine.

Step 1. The first thing that one must know about the system is the working fluid. Knowing the fluid enables us to find the properties of the fluid we need to know to design the system. These properties are as follows:

c_p specific heat at constant pressure Btu/lb °R

μ viscosity (absolute) ν kinematic viscosity

ρ density $\nu = \dfrac{\mu}{\rho}$

C_{HT} heat transfer coefficient

Step 2. From the engine manufacturer or from other sources such as SNAME, the following must be found:

- temperatures
- required pressures
- heat load
- flow rate

Step 3. Find the maximum velocity for a particular fluid in a design text, such as *Marine Engineering,* edited by Roy L. Harrington. This velocity of the fluid is based on empirical data from the design of many fluid systems. If the velocity is too low, the pipe size is large and the piping costs more money. If the velocity is too high, it could cause cavitation, and pipe erosion, or use a large amount of energy to pump the fluid around the system.

Step 4. Now that we have the maximum velocity and the flow rate from Step 2, we can calculate the pipe size.

$$\text{MassFlowRate} = \rho \text{ x Area x Velocity}$$

$$\text{Area} = \pi \times \frac{D^2}{4}$$

From the above, we can calculate the diameter of the pipe. United States Coast Guard (USCG) and American Bureau of Shipping (ABS) provide rules to calculate the wall thickness of the pipe, and choose the pipe schedule—40, 80, or 120.

Step 5. Sketch out the system with a rough idea of the piping lengths. This should include pumps, valves, heat exchangers, and anything else associated with the system. Typical diagrams for marine engineering systems can be found in Harrington and other sources. *Modern Marine Engineer's Manual, Volume II,* edited by E. Hunt, has diagrams especially relevant to diesel engines. Engine manufacturers have project manuals that include all systems for their particular engine.

Step 6. Using the fluid velocity, fluid viscosity, fluid density, and the pipe diameter, the Reynolds number can be calculated.

Using friction loss theory, the piping losses of the system can be calculated. Pressure losses for heat exchangers and other components must be added. Pressure losses due to

$$\mathrm{Re} = \frac{\rho VD}{\mu} \quad \frac{VD}{\nu}$$

gravity must be calculated. The velocity head must be calculated from the velocity difference across the pump. Any pressure requirements of the system by the engine manufacturer must be taken into account. All these losses must be used to find the pressure difference across the pump.

Step 7. Using the mass flow rate and the pressure difference across the pump, the liquid horsepower of the pump can be calculated.

$$\text{Liquid} \quad HP = \frac{\dot{m}}{33000} \frac{\Delta H}{} \quad \frac{\dot{m}}{\Delta H} \quad \text{flow in } \frac{\text{lb}}{\text{min}}$$

head in feet of fluid flowing is the pressure difference across the pump

Step 8. The type of pump used for the system is chosen based on the fluid being pumped. The efficiency of this type of pump can be looked up and used to calculate the input horsepower of the pump. This would be the horsepower required at the shaft of the pump (pump shaft horsepower).

Step 9. The electric horsepower of the motor can then be determined, using the input horsepower of the pump (shaft horsepower) and the efficiency of the motor.

Step 10. Other particular components to the system such as instrumentation must be chosen.

Engine Room Ventilation System

The intake system of diesel engines is provided by the manufacturer. This consists of the filters around the turbocharger, the turbocharger, intercoolers, and the intake manifold.

The marine engineer has to design the engine room ventilation system. This consists of the following:

- supply fans
- exhaust fans
- supply fan air filters
- duct work

Since this is a ventilation system, references dealing with building ventilation systems are a good source of design data.

The amount of air needed must fulfill the requirements of the engine, plus the air that is needed for ventilation. The amount of air required for the engine can be found from the engine manufacturer or from SNAME 3-49 in the form of the air-fuel ratio, BSFC, and the engine horsepower. The air used by the diesel generators, air compressors, and auxiliary boiler also must be taken into account.

Exhaust System

The exhaust system on a diesel engine starts at the outlet of the turbocharger. The engine manufacturer takes care of the exhaust manifold leading from the cylinders to the turbocharger.

The marine engineer must design the system with the following components, depending on how much heat is to be recovered from the exhaust:

1. A silencer is used if no heat is to be recovered.
2. A thermal oil heat exchanger with appropriate piping and pumps is used to extract heat while underway.
3. An economizer is used in conjunction with an auxiliary boiler if the heat is to be extracted and auxiliary steam is to be produced while underway.
4. A large waste heat boiler is used with a turbogenerator system if enough heat is available to supply the ship with electricity while underway.
5. Exhaust piping ending at the stack.

The pressure drop across the whole system is very important. This pressure drop is the back pressure exerted on the outlet of the turbocharger. The engine manufacturer usually gives this back pressure in inches of water. These units are used because very low pressures are needed in the exhaust after the turbocharger. This back pressure is then the total pressure drop from the outlet of the turbine of the turbocharger to the top of the stack. The piping system, with any associated equipment, has a very large diameter to make sure the pressure drop is within specified limits.

The exhaust flow can be obtained from the engine manufacturer or SNAME 3-49. Items needed to calculate exhaust flow include BSFC, air fuel ratio, and engine horsepower. SNAME 3-49 also outlines the steps required to calculate how much steam can be extracted by a waste heat boiler.

The choice of the type of waste heat equipment employed depends on the efficiency of the engine and the horsepower. With high-horsepower, older, less efficient engines, enough waste heat was available to power an electrical power plant. This plant had to be run at very low pressures because of the relatively low temperature exhaust of the diesel engine, which made for a very inefficient steam plant. However, if enough electricity could be generated to satisfy the power requirements at sea, the electricity was essentially free.

Today the engines are more efficient and may not produce enough waste heat to generate quantities of electricity to be economically viable. In these cases a simple fire tube waste heat boiler may be used. In very efficient engines, a thermal oil heat exchanger is used instead of the waste heat boiler. The thermal oil is used for general low temperature heating aboard the ship and in the engine room. Another simple system is to use an oil-fired boiler with an economizer located in the stack. Depending on the engine and horsepower, the waste heat system can have a vast range of complexity.

All these systems require some type of soot blowers. Even though diesel propulsion engines run with clear stacks, they generate enough soot to demand the use of soot blowers. This promotes good heat transfer and eliminates the possibility of stack fires. This is especially true because of the finned tubes associated with any heat exchanger in the stack. Soot blowers on a steamship use steam. A diesel ship uses compressed air because of the plentiful supply of high-pressure starting air.

The intake and exhaust systems are no different than any other fluid flow system onboard a ship. The same equations are used for pressure drop calculations. The only difference is in the working fluid, which is a gas. The viscosity and density at the working temperatures are used just as they are in the calculations for a water system. Since most of the gas in the diesel exhaust is air, the properties of air can be used for a good approximation.

Study Questions

1. The Reynolds number allows one to obtain the:
   A. velocity head
   B. static head
   C. gravity head
   D. friction head
2. A thermal oil system would be a component in which of the following diesel ship systems?
   A. exhaust
   B. fuel oil
   C. lube oil
   D. intake
3. The symbol represents what fluid property?
   A. density
   B. kinematic viscosity
   C. absolute viscosity
   D. specific heat
4. The liquid horsepower of a pump is the horsepower required to drive the:
   A. fluid
   B. pump shaft
   C. coupling
   D. motor

Chapter 7

Fuel Oil Service, Lubrication Oil Service, and Treatment Systems

Fuel Oil Transfer System

The fuel oil transfer system takes Heavy Fuel Oil (HFO) from the double-bottom tanks and deep tanks and transfers it to a settling tank. Incorporated into this system is a port and starboard filling station that is used to fill all fuel tanks.

The components needed in this HFO system are:

1. Double-bottom tanks with vents and sounding tubes.
2. Deep tanks with vents and sounding tubes.
3. Piping.
4. Suction manifold.
5. Discharge manifold.
6. Suction strainer for the pump.
7. Transfer pump-electrically driven rotary pump.
8. Settling tank; Heavy oil (settling tank has a sloped bottom and drain for the removal of water).
9. Steam-heating coils for all tanks.
10. Tank level indicators.
11. Appropriate pressure and temperature indicators.

Also included is a distillate fuel oil system to pump up the distillate oil day tank. This system utilizes No. 2 fuel oil that requires no heat.

The mass flow rate of the fuel in this system depends on how fast you want to transfer the oil from the double-bottom tanks to the settling tanks. This usually is done during one four-hour watch. Also entering in to this calculation is the BSFC of the main engine and the brake horsepower. The amount of fuel the generators and auxiliary boiler use in twenty four hours must be taken into account.

It is important to take temperature into consideration, since we are pumping oil and its viscosity and density varies with temperature.

When filling the fuel tanks, the pressure of normal bunkering facilities must be used to size the shore connections and the fill lines to the tanks.

Ships burning heavy oil use fuel oil purifiers to pump the oil from the settling tanks to the day tanks. The purifier is piped to the day tank with an overflow back to the settler. The size of the purifier depends upon the BSFC and brake horsepower (BHP) of the main engine, and the fuel flow of the generators and the auxiliary boiler. Ships have a distillate diesel oil system and a HFO system. The distillate diesel oil system, oil is pumped from the double bottoms directly to the day tanks by a small rotary pump or a small purifier.

Fuel Oil Service System

The fuel oil service system takes oil from the day or service tank and supplies it to the engine at the proper viscosity, pressure, and purity.

Components of the system:

1. Diesel oil service tank-no steam heating-sloped bottom and drain for water removal.
2. Heavy oil service tank-steam heating required-sloped bottom and drain for water removal.
3. Three-way change over valve between diesel oil and HFO.
4. Mixing tank with steam heating, oil vapor condensation trap, and vent.
5. Suction strainers for fuel oil rotary booster pumps.
6. Two fuel oil booster pumps with pressure-sensing motor controllers to automatically switch pumps when one fails. Isolation valves and check valve are needed on the discharge to prevent backflow.
7. Two fuel oil heaters.
8. Some type of final filter after the fuel oil heaters and before the main engine. On modern diesel ships usually a very fine self-indexing back-flushing filter is used.
9. Viscometer. This device measures viscosity and controls the steam flow to the fuel oil heaters to maintain the proper viscosity for fuel injection.
10. Meters to measure the amount of fuel consumed.
11. Pressure regulator. Controls the pressure in the high-pressure fuel injection pumps by throttling the excess flow of oil back to the mixing tank from the engine in older ships. Recent systems and all new construction use a pressurized mixing tank. HFO is pumped from the day tank to the recirculation loop by fuel oil feed pumps. The pressure in the recirculation loop is controlled by a pressure regulator that discharges back to the suction of the feed pump.
12. Two fuel oil feed pumps with pressure-sensing motor controllers to automatically switch pumps when one fails. Isolation valves and check valve are needed on the dis charge to prevent backflow.
13. Tank level indicators and alarms.
14. Appropriate pressure gauges and thermometers.
15. Essentially the fuel flows from the service tanks by gravity or by fuel oil feed pump to the mixing tank. The fuel oil booster pumps circulate the oil in a loop from the mixing tank through the fuel oil heaters to the high-pressure injection pumps and back to the mixing tank. Fuel is added from the service tanks as the engine burns the fuel. The recirculated flow in the loop is two to three times the flow needed to supply the engine at maximum continuous rating. This is to keep the oil at the proper viscosity.

There is a diesel oil service tank and a heavy oil service tank. Many modern slow-speed engines can run completely on heavy oil. During shut down and repairs, diesel oil is needed in the lines to prevent fuel solidification during shut down.

All service day and settling tanks are higher than they are wide. They also have a sloped bottom and a drain. This is to facilitate the removal of sludge and water. They also have a high and low suction to prevent sludge and water from getting into the fuel system.

Steam tracing lines are incorporated into the insulation of the fuel oil service piping to ensure the fuel does not solidify.

The mixing tank is pressurized in more modern systems because of the high temperatures that today's heavy fuel has to be heated to. If the mixing tank is not pressurized and vented through a special condensation trap, the lighter constituents of the fuel boils off and condenses in the vent line. This condensation would appear to be fuel pouring out of the vent.

The pressure and temperature at the high-pressure injection pumps is specified by the engine manufacturer. The fuel flow to the engine also is specified. The BSFC and BHP can be used to get the fuel flow needed by the engine. Remember, the recirculated flow will be two to three times this value.

Lubrication System

The lubrication system takes oil from the engine sump. This sump is incorporated into the crankcase for small engines and located in the double-bottom below the engine for main propulsion engines. A pipe connecting the lower part of the engine to the sump in the double-bottom tank is required. The oil is pumped through the lube oil coolers or is bypassed around the coolers by a thermostatic valve to maintain a prescribed temperature. The oil is then delivered to a lube oil header that supplies the bearings with pressurized cooled oil. The oil then flows out the sides of the journal bearings and back into the sump.

Components of the lubrication system are:

1. Lube oil sump.
2. Lube oil service pumps. Two rotary pumps with oil pressure sensing automatic change-over in case of pump failure. All pumps using this parallel standby configuration will have a check valve in each of the pump discharges to prevent backflow through the idle pump. All slow-speed engines use electrically driven pumps. Medium-speed engines may use a direct drive pump and an electrically driven pump as automatic standby.
3. Suction strainers for the pumps.
4. Discharge filter located before the oil coolers, possibly a self-indexing, back-flushing filter.
5. Pressure regulator. This will be on the discharge of the pumps and lead back to the suction to regulate the oil pressure in the system.
6. Oil coolers. Two oil coolers are used, one being for standby.
7. Thermostatic valve. Controls oil temperature within specifications.
8. Pressure gauges, thermometers, and alarms.
9. Level indicator and alarms for the sump.
10. A heater may be installed in the bypass line around the coolers. This is used to heat the lube oil of a cold engine.

Various versions of the lube oil system are available, depending on the use of the discharge filter.

Full Flow or Continuous. In this system all the oil goes through the filter. The finer the filter, the larger the pressure drop across the filter unless the filter is made larger and, thus, more expensive. There are several ways around this, as seen with the shunt and by pass system.

Shunt System. A smaller, finer filter is used with a shunt around the full flow filter. A pressure regulator or orifice is used to control the flow through the shunt so that a certain percentage goes through the shunt and a certain percentage goes through the filter.

Bypass System. In this system some oil is taken from the discharge of the pump and run through a small fine bypass filter back to the sump. The amount of oil bypassed through this filter and back to the sump is controlled by a pressure regulator or orifice.

Batch System. Oil is taken out of the sump and pumped into a separate tank. A separate system is used to purify or filter the oil.

Sump or Continuous Circulation System. This system uses a completely separate pump and filter to work on the oil in the sump. Most modern engines use a centrifugal purifier instead of a filter for this type of system.

Most main propulsion engines use the full flow (continuous) system in conjunction with a sump (continuous circulation) system using a purifier.

The lube oil storage tanks and a lube oil settler is located above the main engine, with a shore connection to the storage tank. The settler has steam heating coils. It connects to the discharge of the main oil pumps or has a separate pump for pumping sump oil to the settler. This settler is only used as part of a batch system when the main sump is contaminated. These tanks are located at an elevation above the main engine, so that oil can be gravitated to the engine sump.

This is one system where the mass flow rate, temperature, heat rejected, and required pressure at the main engine must be obtained from the engine manufacturer. For a preliminary design these items can be taken from existing plants, but the final design must meet the engine manufacturer's requirements. The engine manufacturer also will specify the type of oil.

The above is the basic lubrication oil system all engines require. Different manufacturers may require additional systems. For example, Sulzer requires an additional set of crosshead lube oil pumps for its slow-speed engines. These take suction from the main lube oil system pump discharge and elevate the pressure and supply this higher-pressure oil through the articulated oil line attached to the crosshead bearing. MAN B&W require a completely separate camshaft lubricating oil system for its slow-speed engines. All slow-speed engines have a separate cylinder lubrication system. Once through, cylinder oil is fed to lubricating quills screwed into the cylinder liner walls by a set of small, positive displacement piston pumps. Colt-Pielstick medium-speed engines require a separate rocker arm lube oil system.

Treatment Systems

Centrifuges or Purifiers. These devices are used in both the lube oil and fuel oil purifications systems. They remove water and particulate matter. They cannot remove dissolved material from oil.

The separating force in a centrifugal separator depends upon:

1. Viscosity of the oil. The lower the viscosity, the easier it is to separate impurities from the oil. That is why the purifier system has a heater and the oil is heated to about 190

°F. The oil is not heated above this temperature because that would cause any moisture in the oil to boil.

2. Difference in specific gravity. The greater the difference in specific gravity between water and oil, the easier it is to separate the impurities from the oil. Heating the oil lowers its specific gravity. Heating water and solid impurities does the same, but to a lesser degree. The overall effect of heating the oil and impurity mixture is to amplify the differences in specific gravity.
3. Centrifugal force.

$$a_n = \frac{V_t^2}{r} = \omega^2 r \qquad \text{since} \qquad V_t = r\omega$$

   Since centrifugal force equals mass times normal acceleration, it can be seen from above formula that the higher the rpm of the purifier, and the farther the distance of the oil from the centerline of the rotating bowl, the higher the separating force.
4. Time. The lower the throughput of the purifier, the longer the time the oil and impurity mixture remains under the centrifugal force and more impurities can be removed. The designer of a purifier system must pay great attention to the throughput. If a nominal throughput is given, then the actual throughput must be a small fraction of the nominal throughput if very dirty oil is being purified. The more impurities in the oil, whether fuel or lubricating, the lower the throughput must be.
5. Distance impurities travel before being removed. The shorter the distance the impurities travel before they strike a removing component, the faster they are removed from the mixture. Most modern purifiers have a disk stack. As soon as an impurity hits a disk it is removed from the mixture. The disk stack also accelerates the oil to the rotating speed necessary for the centrifugal force to work.

Purifiers have several modes of operation:

- Separator. This set-up is mainly to extract water, with some sludge being removed.
- Clarifier. This mode removes only solid impurities. Water will cause the bowl to completely fill with water.
- Purifier. This mode is similar to the separator, but the system is set-up to remove both water and solid materials. Both the separator and the purifier require a water seal to be established before oil can be admitted.

Figures 7.1 and 7.2 represent half the bowl of a separator and a clarifier. The main difference between the clarifier and the separator is that the clarifier does not have a water outlet. The inner diameter of the gravity disk is the same diameter as the clean oil outlet tube.

At one time, the only way to clean these machines was to take them apart. On a heavy-oil-burning diesel ship the fuel oil purifier would have had to be cleaned hourly. The lube oil purifier had the same problem.

A new self-cleaning purifier was created. To have this type of self-cleaning purifier the side of the bowl was changed to a wedge shape. This concentrates the sludge in a smaller volume. With this area of concentrated sludge, a movable bowl bottom can momentarily drop, causing the sludge to shoot out of the circumference of the bowl. The movable bowl

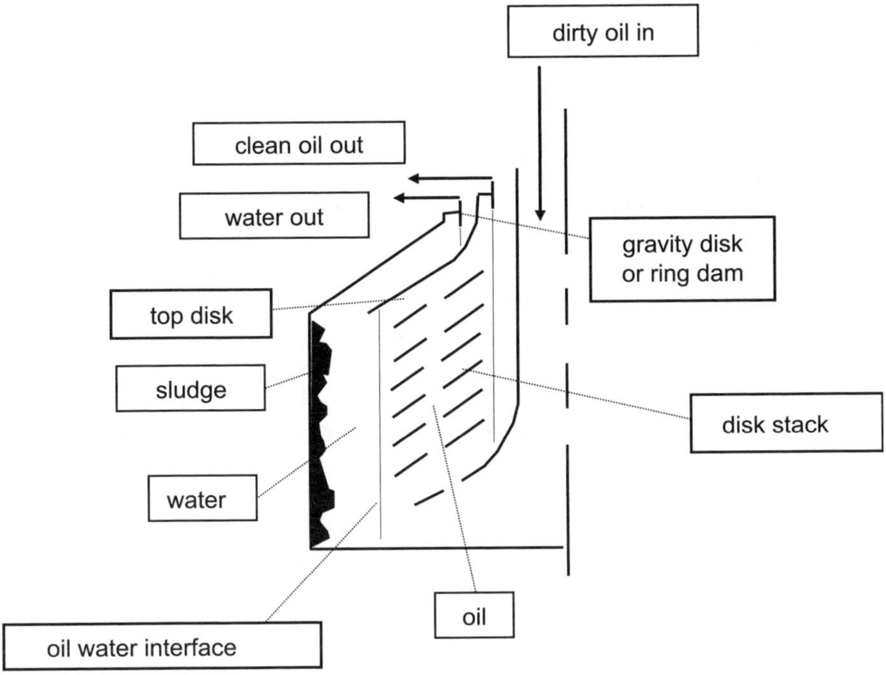

Fig. 7.1. Centrifugal separator or purifier set-up

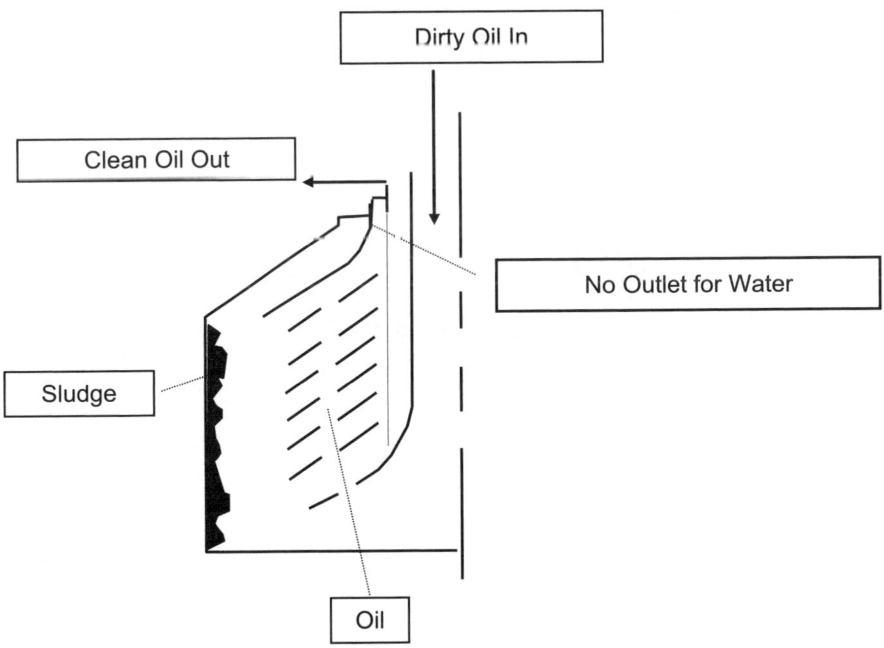

Fig. 7.2. Centrifugal clarifier setup

bottom is held up by water in the hydraulic chamber (fig. 7.3). A computer controls the sequence of events to shoot the bowl, which are:

1. The oil inlet is shut off.
2. Extra water is put into the bowl if running as a separator or purifier. This is called purging.
3. Water is drained from the hydraulic chamber.
4. The bowl bottom drops and shoots out the sludge with a minimum of oil.
5. Water is replaced in the hydraulic chamber. The water in the hydraulic chamber is drained and replaced in a fraction of a second. This means that the sludge is shot out of the purifier in a fraction of a second.
6. The water seal is reestablished if running as a separator or purifier.
7. Oil flow is restored.

The computer has an internal clock that can be set. The dirtier the oil, the shorter the time interval for shooting the bowl. Approximately every thirty days the purifier has to be manually taken apart and cleaned.

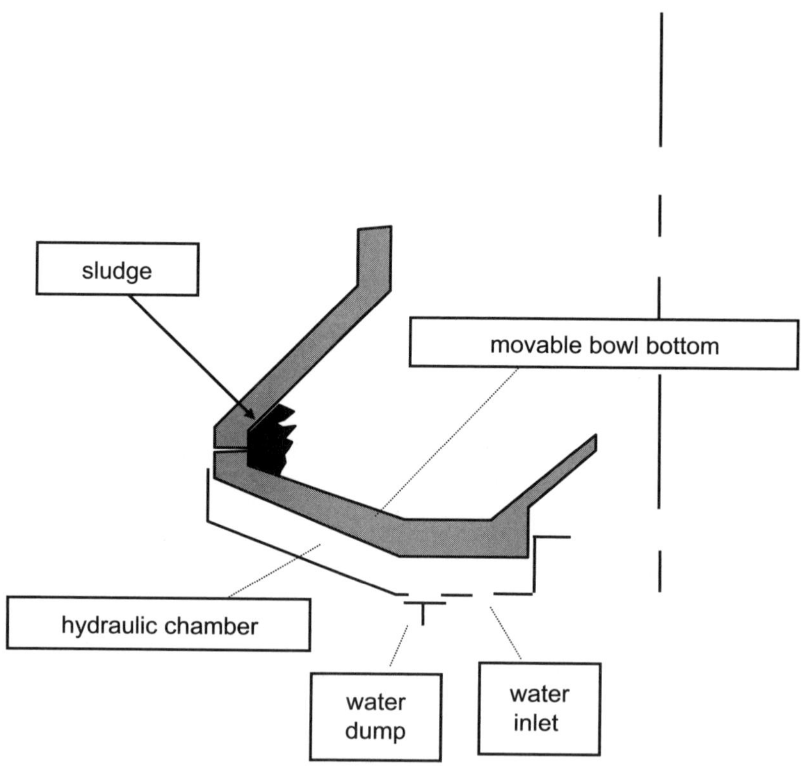

Fig. 7.3. Centrifuge wedge shape and movable bowl bottom

Another feature added to the self-cleaning purifier was the paring disk at the clean oil and water outlets. This is essentially a diffuser of a centrifugal pump. Since the outlet fluids are already spinning, these paring disks just have to convert the kinetic energy to pressure energy and send these fluids to the desired location. These paring chambers act like discharge pumps of the purifiers. The older nonself-cleaning purifiers collect the clean oil by gravity and use a discharge pump for the clean oil, and an inlet pump for the incoming dirty oil.

Controlling the oil water interface during operation is critical to the purifier. The interface should be as near to the periphery of the bowl as possible to increase the centrifugal force on the dirty oil. If the interface is too far out, the oil will break over and mix with the water leaving the purifier.

Factors Controlling the Oil-Water Interface

1. The higher the viscosity and density of the oil being purified, the closer the interface will be to the periphery of the bowl. This is true because the heavier the oil, the more centrifugal force is exerted forcing the interface out. The more viscous the oil, the more resistance to flow out of the purifier. This causes the interface to move out.
2. The higher the throughput, the higher the internal friction of the oil leaving the purifier, and the interface moves out.
3. The higher the back pressure of the clean oil leaving the purifier, the more the interface will move out. Modern purifiers have a back pressure valve for this reason.
4. The larger the internal diameter of the gravity disk or ring dam, the farther out the interface moves.

The internal diameter of the gravity disk is selected based on the density and viscosity of the oil being purified. It is chosen so that we have no break over of the oil into the water. The back pressure valve is used to fine-tune the system to get the oil-water interface as far out as possible without breaking over.

This problem of oil-water interface and a water seal can be eliminated by using the Alcap system manufactured by DeLaval. This system is a clarifier without a water seal. Located in the clean oil outlet is a sensitive moisture sensor, which will shoot the bowl if it detects water. This eliminates the water that collects inside the bowl of the clarifier. Remember, a clarifier does not have a ordinarily water outlet.

Series and Parallel Operation

When two purifiers run in parallel, the throughput through each purifier is cut in half. This allows smaller particles to be removed from the oil. If the purifiers are in series, the second can back up the first in case of a malfunction. The clarifier must be placed after a separator or purifier, since the clarifier will fill with water if any water is present in the oil.

Study Questions

1. What moves the fuel from the double bottoms to the settler?
   A. rotary pump
   B. gravity
   C. purifier
   D. diaphragm pump
2. What controls the viscosity of the heavy fuel leaving the fuel oil heater of a modern diesel ship?
   A. temperature regulator
   B. viscometer
   C. gravimeter
   D. lubrimeter
3. Which lubricating oil system has all the oil going to the engine through the filter?
   A. continuous
   B. shunt
   C. sump
   D. bypass
4. Which lubricating oil system diverts part of the oil flow around the main filter?
   A. continuous
   B. shunt
   C. sump
   D. bypass
5. In diagram A, which number represents the oil-water interface?
   A. 3
   B. 4
   C. 5
   D. 6

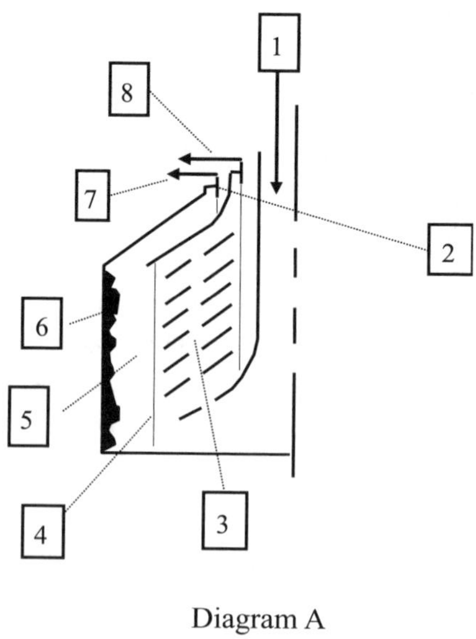

Diagram A

6. In diagram A, which number represents the ring dam location?
   A. 2
   B. 3
   C. 7
   D. 8
7. The device in diagram A represents a:
   A. separator
   B. clarifier
   C. both
   D. neither
8. What moves the heavy fuel from the settler to the service tank?
   A. rotary pump
   B. gravity
   C. purifier
   D. diaphragm pump
9. A pressurized mixing tank is used in what system?
   A. lube oil system
   B. distillate fuel oil service system
   C. heavy fuel oil service system
   D. heavy fuel oil purifier system
10. Which lubricating oil system has part of the oil flow not entering the engine?
    A. continuous
    B. shunt
    C. full flow
    D. bypass
11. In which type lubricating oil system is the purifier located?
    A. continuous
    B. shunt
    C. sump
    D. bypass
12. In diagram A, which number represents particulate matter?
    A. 3
    B. 4
    C. 5
    D. 6
13. Which diameter of a gravity disk controls the oil-water interface?
    A. outer
    B. inner
    C. both
    D. neither

14. What keeps the bowl closed in a self-cleaning purifier?
    A. fuel oil
    B. lube oil
    C. springs
    D. water

Chapter 8
Cooling Water Systems

Jacket Water System

The primary coolant in the main engine and diesel generators is a freshwater system that is a closed loop. The basic components of the system are:

1. Two centrifugal circulating pumps. These are arranged in parallel with an automatic changeover feature incorporated into the motor controller circuit. The changeover is based on discharge pressure or flow. These pumps are electric-motor driven, centrifugal pumps. All slow-speed engines use electrically driven pumps. Medium-speed engines may use a direct-drive pump and an electrically driven pump as automatic standby. A check valve must be placed on the discharge of each centrifugal pump to prevent back-flow during automatic changeover.
2. Two coolers. Each heat exchanger can take all the heat from the primary coolant.
3. Temperature regulating valve. This three-way valve sends the primary coolant either around the heat exchanger (too cold) or through it (too hot), based on the cooling temperature. This keeps the coolant moving to the engine at the prescribed level.
4. Head tank. This tank, located high in the engine room, provides a place for the expansion of the primary cooling water. A gauge glass, high- and low-level alarms, and an air vent are provided. Make-up water is added to this tank through a funnel and pipe arrangement. The potable water system and any other system on the ship must have an air gap between them to prevent contamination of the drinking water. Just about all oceangoing ships use distilled potable water for make-up. Chemicals are added in this tank on some ships, while others use a dosing tank.

This system supplies water to the bottom of the cylinder liner. The water then goes through the cylinder head and back into the system. This system has the major components listed above, plus the following:

1. Distiller. The distiller takes heat from the main engine jacket water and produces fresh water.
2. Jacket water heater. This heater raises the temperature of the jacket water before start-up. This heater is located on the bypass line of the three-way temperature control valve that goes around the coolers.
3. Flow stabilizer. This device, located after the engine heads, vents any air in the coolant to the head tank and stabilizes the flow down-stream to the circulating pumps.
4. Vent lines. These small lines run from the cylinder head and the pumps and vent any air in the system to the head tank.

To design this system, you must know the mass flow rate, temperature of coolant to the engine, coolant pressure at the engine inlet, and the heat rejected by the engine to the coolant. This information can be obtained from the engine manufacturer. For an approximation, the following can be done. An estimate of the heat rejected by the engine to the coolant can be obtained from SNAME 3-49. Temperatures around the heat exchanger and the engine can be taken from typical plants. Using the specific heat of water and the heat exchanger equations, mass flows can be calculated.

$$Q = \dot{m} c_p (\Delta T) \qquad \text{Units of Q Btu/min}$$

$$\text{Q} = \text{Coef (Area) (LMTD)}$$

LMTD is the Log Mean Temperature Difference. This is a representation of the temperature difference of the fluids in the heat exchanger.

The heat transfer area of the heat exchanger is one of the final parameters that needs to be calculated. This methodology can be used to design any cooling water system the engine requires.

Other freshwater cooling systems that may be required are:

1. Piston cooling water system. This system is used on loop-scavenged engines and older uniflow engines. The uniflow engines are no longer being built with this system. It is a typical freshwater system with a baffled, vented tank located in the lower part of the engine room. This takes the place of the head tank. Also incorporated in this tank is an oil separator. This device is needed because the telescopic tubes always entrain some oil into the piston cooling water. This tank contains a heat exchanger for cooling and another for heating during engine warm-up.

$$LMTD = \frac{\Delta t_a - \Delta t_b}{\ln \dfrac{\Delta t_a}{\Delta t_b}}$$

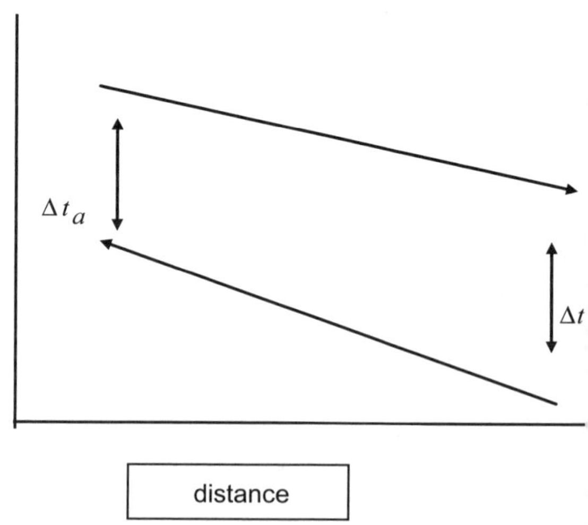

Fig. 8.1. Temperature versus distance in a counter flow heat exchanger

2. Injector cooling water. This system is used on some engines to cool the injectors during operation and heat them during start-up. Most modern engines have eliminated this system by using better cooling design in the head near the injector.
3. Cooling for diesel generators. A miniature version of the cooling and lube oil systems are needed for the generators. They are mounted on the skid supplied with the generator.
4. Air conditioning condenser cooling.
5. Starting air compressor cooling.
6. Charge air cooling.

These systems can be cooled with saltwater. This means we have to use expensive corrosion-resistant piping to supply saltwater to all heat exchangers mentioned above. The present day trend is to use another closed loop freshwater system to cool all the above heat exchangers. The saltwater system will then be centrally located and cool two jacket water coolers and two central coolers.

Saltwater Cooling System

On modern diesel ships, this system is centrally located and compact, so as to minimize the amount of expensive corrosion-resistant material, such as copper-nickel and monel. This is an open system. The following components are usually used:

1. Centrifugal circulating pumps. There usually are three of these pumps. Two are sized to carry the full sea load, with automatic changeover due to pressure or flow loss. One usually is a smaller pump for in-port use. A check valve must be placed on the discharge of each centrifugal pump to prevent backflow during automatic changeover.
2. Sea chests both port and starboard, and high and low, with steam and compressed air lines to blow out debris.
3. Overboard discharge.
4. Suction strainers for the pumps.
5. Chloropac Unit. This unit uses saltwater and electricity to produce chlorine, acting as a biocide inside the saltwater components.
6. Bilge injection valve. This valve is required on the largest saltwater pump in the engine room. This usually is the saltwater circulating system, but it could be the ballast pump. The bilge injection valve pumps out the engine room during a large casualty. In a modern saltwater-circulating system, a temperature regulating valve is used to divert some of the water to the suction side of the circulating pump before it is discharged overboard. This is to ensure that the circulating water does not get too cold when sailing in cold climates.
7. Scoop injection. This eliminates the use of a circulating pump while underway.
8. Since this is an open system, usually completely submerged in saltwater, the circulating pump only has to overcome the friction head loss and any velocity head loss. If the overboard discharge rises above the surface of the saltwater, then a gravity head loss must be added. The mass flow of the saltwater can be calculated knowing the total heat that must be rejected and the usual temperatures involved. Engine manufacturers also supply flow rate and pressures required for their engine systems.

The saltwater circulating system and the central freshwater cooling system usually have heat exchangers connected in parallel. When this is done, it is possible that one of the heat exchangers can take most of the water. The heat exchanger with the lowest resistance will take more water. To compensate, an orifice may have to be placed before the heat exchangers with lower resistance to adjust the flow. The flow must be adjusted so that each heat exchanger absorbs the heat flow required.

Several types of heat exchangers are used in diesel engine systems.

1. Tube and shell. Saltwater usually is circulated in the tubes to promote easy cleaning. This type of heat exchanger is more difficult and expensive to manufacture. It is very well suited to exchange heat between high-pressure liquids.
2. Plate type heat exchanger. This heat exchanger is suited to exchange heat between low-pressure liquids. This is what we have in diesel engine fluid systems. It is easier to manufacture because the plates can be stamped out.

The fluid flows through the diagonal holes that have a circular gasket around them or the fluid flows across a plate that has the large gasket enclosing the other two diagonal holes. With mirror-image gasket patterns alternating in the plate stack, one fluid flows on one side of the plates and another fluid flows on the other side. With the appropriate end plates holding the plate stack together, and the appropriate flanges over the holes in the plates, we have our plate-type heat exchanger. The plates can be stamped out with ripples incorporated in the plate to provide better heat transfer by inducing more turbulence. However, this also produces a larger pressure drop across the heat exchanger.

1. Keel cooler. This is a bundle of tubes located parallel to the keel on the outside of the hull. Freshwater flows through the tubes, and the outside of the tube bundle is submerged in saltwater. The motion of the ship causes the saltwater to flow over the tubes. This type of heat exchanger is used on smaller craft up to the size of tugboats.
2. Radiator. This liquid-to-air heat exchanger is used on emergency diesel generator engines so that the cooling can be completely independent of the ship systems during emergencies.

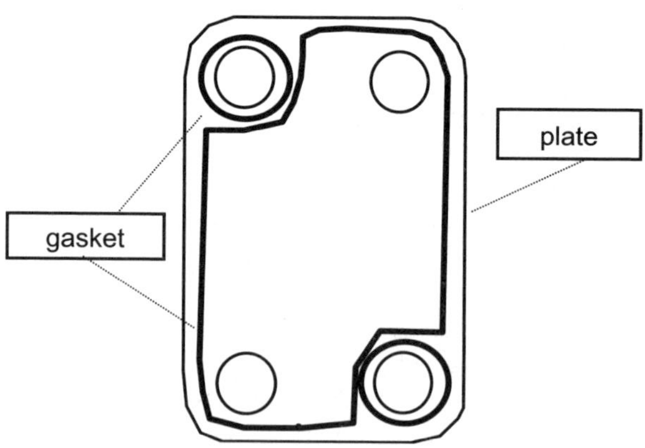

Fig. 8.2. Plate of a plate-type heat exchanger

Freshwater System Maintenance and Treatment

The fluid in the freshwater system must be distilled or demineralized by a zeolite bed to produce freshwater. Minerals create scale on the surfaces of the system when the temperature becomes too hot. Calcium carbonate, a major component of scale, comes out of solution when the temperature goes up. If scale forms, it insulates the surface and raise the temperature even more, promoting more scale. This can cause even more problems in the cylinder liner where this scale-induced hot spot could cause lubricating oil film failure on the other side of the liner. The saltwater side of a heat exchanger also will develop deposits from biological action and sediment.

The saltwater side can be cleaned mechanically. The freshwater side can be cleaned mechanically in certain areas, such as the heat exchanger. However, inside the engine this is not possible. Acid cleaning with a product, such as safe acid, usually is done where the surfaces are inaccessible. The color of the acid goes from pink to brown or blue to green as the acid is used up.

Chemicals are aided to control corrosion and erosion on the freshwater side. Corrosion is due to oxygen being dissolved in the water. Erosion is due to fluid flow and cavitation. The cavitation is evident especially in the liners of trunk piston engines. The liner of these engines must absorb side thrust from the piston during combustion. This sets up vibrations in the liner that cause cavitation on its waterside surface. The cavitation forms explosive bubbles of water vapor. These micro explosions tear off pieces of liner and form pits.

The chemical treatment system contains an alkaline chemical to control corrosion and a plating agent to control erosion. These are added to the freshwater side.

A cooling system treatment program must meet four conditions:

1. The system must be clean to start with, so the inhibitors can reach the metal surface.
2. The coolant must be checked for alkalinity (pH), inhibitors, and chlorides. The inhibitors and alkalinity must be maintained at the correct level. The chlorides are checked to see if saltwater is entering the system.
3. The basic coolant must be clean, mineral-free freshwater, containing a minimum of dissolved gases. This is usually distilled water.
4. The system must be mechanically tight. No leaks.
5. There are two basic types of chemical treatment plans, chromate-alkaline and nitrate-borate.

Chromate-Alkaline

This program uses sodium dichromate, which in solution changes to sodium chromate. This is a plating agent that coats the metal surface on a molecular level to protect it against erosion.

Sodium hydroxide is used to get the pH above seven. As in a boiler, a pH slightly above seven will prevent red iron oxide from forming on iron surfaces, preventing corrosion. Black iron oxide forms under these conditions. Black iron oxide is dense and tough. It will protect against more corrosion.

Ameroid (Drew-Ashland) recommends 1,000-3,000 ppm sodium chromate.

B&W recommends a pH of 9.0-9.2. If the pH reaches 9.8, the system should be drained, refilled, and retreated. Too high a pH causes caustic corrosion.

This is a very effective chemical treatment system, but it has one problem. The sodium chromate is very toxic to humans and the environment if the cooling system contents are dumped overboard. In the jacket water system, where the distiller uses the jacket water as a heat source, this toxic chemical treatment could cause a problem.

Nitrate-Borate

Sodium nitrate is the plating agent in this system. Sodium borate is the alkalinity agent in this system. This is a premixed product that should be kept at a level of 3,000-4,500 ppm.

This system is not poisonous, but it can act as a fertilizer. If small quantities of coolant with this treatment enter the lubricating oil, microbes can grow in the oil. Their waste products usually are acid, which attacks the bearings in the crankcase. This happens when a ship enters the hot climate of the Persian Gulf. This is the area where high-temperature-loving microbes live. The purifier usually will heat the oil to a sustained, sufficiently high temperature to kill these microbes. Sometimes the purifier piping system must be modified to keep the oil heated for a sufficient time. Biocide additives also are available for oil.

Study Questions

1. The temperature regulating valve in the jacket water system of a main propulsion engine:
   A. blocks water to the engine if the water temperature is too cold
   B. sends more water to the engine if the temperature is too hot
   C. diverts more water to the cooler if the temperature is too cold
   D. diverts more water to the cooler if the temperature is too hot
2. The log mean temperature difference models the temperature difference in a heat exchanger as it varies with:
   A. pressure
   B. temperature
   C. distance
   D. velocity
3. The saltwater cooling system in a diesel ship usually has how many circulating pumps?
   A. one
   B. two
   C. three
   D. four
4. Which type of heat exchanger is easier to manufacture and is found on diesel merchant vessels?
   A. tube and shell
   B. plate type
   C. radiator
   D. keel cooler
5. Which type of heat exchanger is cooled by saltwater but does not need a saltwater pump?
   A. tube and shell
   B. plate type
   C. radiator
   D. keel cooler
6. What can adjust the pH in a freshwater system?
   A. sodium chloride
   B. sodium chromate
   C. sodium borate
   D. sodium carbide
7. In a diesel ship, the distilling plant takes heat from the:
   A. jacket water system
   B. sea water system
   C. exhaust system
   D. lube oil system

8. Which modern diesel engine must use a piston cooling water system?
   A. four-stroke cycle
   B. cross flow scavenged
   C. loop scavenged
   D. uniflow scavenged

9. The saltwater system has a gravity head if the discharge is:
   A. below the waterline
   B above the waterline
   C. at the waterline
   D. A and C

10. Which type of heat exchanger does not use a liquid?
    A. tube and shell
    B. plate type
    C. radiator
    D. keel cooler

11. What is scale?
    A. calcium carbonate
    B. calcium carbide
    C. sodium carbonate
    D. sodium carbide

12. What can act has a plating agent in a fresh water system?
    A. sodium chloride
    B. sodium chromate
    C. sodium borate
    D. sodium carbide

Chapter 9
Starting and Reversing Systems

Small Engine Starting Systems

Small engines use some type of small starting motor. This motor must have a high starting torque to overcome static friction.

1. Electric motor. A series-wound DC motor is used because of the high-starting torque it produces. The field and armature are in series. Some recent small starting motors use permanent magnets as the field. The source of the electricity is a lead acid battery. The current is supplied from the battery through the use of a solenoid. This is a remote switch commonly called a relay. The motor engages the engine through the use of a pinion on the starting motor and a ring gear on the engine flywheel. A Bendix or Dyer drive will engage and disengage the starting motor from the ring gear. Both of these devices are used in conjunction with an overriding clutch. The Bendix drive uses a threaded drive shaft with an internally threaded pinion. The inertia of the pinion helps to screw the pinion into the ring gear. The starting of the engine screws the pinion back towards the starting motor, disengaging it. The Dyer drive is similar but it uses a lever and a solenoid to engage the pinion.

   All small starting motors use some type of similar device to engage and disengage the starting motor.
2. Hydraulic motor. This system uses a hydraulic motor to start the engine. Hydraulic energy is stored in accumulators. These are high-pressure cylinders precharged with nitrogen. They have a piston between the nitrogen and the oil to prevent nitrogen from dissolving in the oil. The pressure in this system can be as high as 3,000 psi. To obtain this pressure, a pump must be provided. This pump can be electrically driven, as in the case of emergency diesel generators. It can be an attached pump. In this case, a special pressure sensing valve must be able to switch the discharge of the pump from the accumulators to the oil storage tank. A simple relief valve will not do, because all the energy supplied to the oil to pop the relief valve will go into heat. The system has a control valve to allow the high pressure oil to go from the accumulators to the hydraulic motor to start the diesel engine and back to the oil storage tank. This system usually has a hand pump to charge the accumulators when no other power is available.
3. Air motor. This system uses an air motor with air supplied from compressed air tanks. A compressor is used to pump up the tanks. Several types of air motors are available, including piston, vane, and turbine. All air motors, except the turbine, need a drip oiler at the air inlet to oil the motor for lubrication.

These auxiliary starting motors are used with high-speed engines and possibly smaller, medium-speed engines. For engines larger than this, direct air starting is used. Compressed air is admitted during the power stroke through timed air start valves in the head of the engine. The compressed air pushes on the piston, spinning the engine for starting.

Starting Compressed Air Storage System

A large compressed air system is needed to supply the starting air. Because of the large amount of air needed, the air is stored at 500-600 psi. The compressed air system has the following components:

1. High-pressure compressors. These compressors are multistage, water-cooled compressors. There are at least two main compressors. A small electrically driven compressor can be run by the emergency diesel generator during cold ship start-up.
2. Air storage tanks. These are large, high-pressure vessels and are subject to the regulations governing unfired pressure vessels by the USCG.
3. Reducing stations. These are needed to reduce the pressure in the storage tanks to levels for the main engine and other ship's services. The main starting compressed air system can be cross connected to the control air and ship's service air through reducing stations. The control air and the ship's service air are miniature lower pressure versions of the main starting air system. The control air system must have dry, oil-free air, otherwise the fine automatic valves and other components will gum up. To prevent this, a refrigerated dryer usually is installed in the control air system. These systems normally are isolated from each other so that a casualty in one does not bring down all the air systems.

Starting air system storage tank facts:

1. 500-600 psi—working pressure to hold more air.
2. Holds twelve starts without the use of the compressor for a direct-drive main propulsion engine. Holds six starts for a marine engine with a controllable reversible pitch propeller.
3. Rule of thumb: The physical volume of the storage tanks is thirty times the physical volume of one engine cylinder.
4. Since the tanks are unfired pressure vessels, they must have a relief valve, a pressure gauge, and a drain.
5. A separate valved line to the tank and a separate valved line from the tank must be provided to ensure moisture does not get into the downstream components.
6. Reducing Stations. Reducing stations are made up of a reducing valve, isolating valves, bypass, pressure gauge, and a relief valve.

The relief valve is especially important, since the piping downstream of the reducing valve is designed for the reduced pressure.

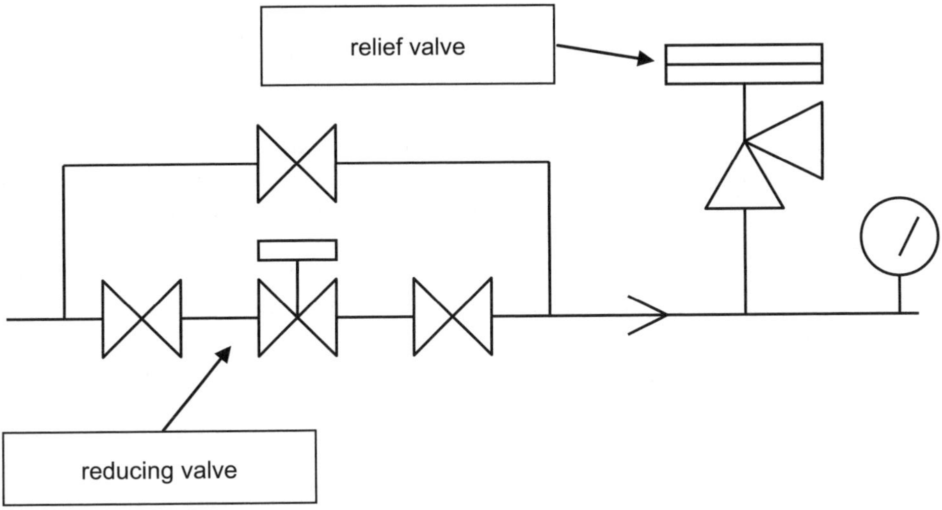

Fig. 9.1. Pressure reducing station

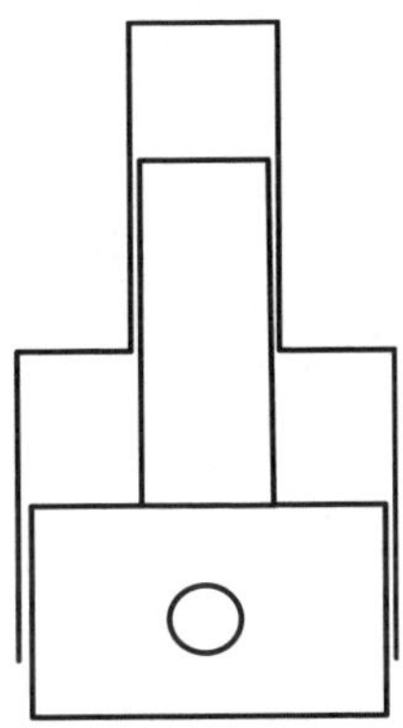

Fig. 9.2. Differential piston

Main Engine Starting Air Compressor

The main engine starting air compressor is a multistage, water-cooled, pressure-lubricated air compressor. The compressor shown has a W configuration. It is two-stage with an intercooler. Since the pressures are so high in this system, multi-staging with intercoolers increases the efficiency of the compressor. The heat of compression can be rejected in the intercooler between stages, making it easier to compress the air. These intercoolers can be air cooled or water cooled on large compressors. They also have a moisture separator, drain, pressure gauge, and a relief valve.

Some compressors use differential pistons and cylinders to obtain two stages with one piston.

On large compressors, a pressurized lubricating oil system, much like the system found on a small diesel engine, is used. On very small single-stage air compressors, a projection on the crankshaft will fling up oil to lubricate the piston and bearings.

One of the more common problems with air compressors is carbon build-up on the suction and discharge valves. Since these valves are gas-operated check valves in cages on the top of the heads, it is a simple matter to remove them. They can be cleaned with kerosene. The kerosene must be blown off the valve parts before reassembly or dieselization of the compressor will occur on start-up.

During operation, the compressor intercoolers should be drained periodically if it does not drain automatically. The relief valves also should be hand lifted periodically to check their operation.

Large compressors have unloaders to limit starting torque. These are spring-loaded oil-operated pistons that hold the suction valves off their seats by spring pressure and are retracted by system oil pressure. Since it has no oil pressure before start-up, the compressor is unloaded. As the compressor spins and builds up oil pressure, the suction valves will seat and the compressor will start to pump air. The mechanism for the delay of the oil pressure to the unloader can be as simple as an orifice or as complex as a solenoids valve controlled by the motor controller. The motor controller will sense when the starting current has decreased and open the solenoid, allowing the oil to load the compressor.

Troubleshooting During Compressor Operation

1. Broken valve. Loss of capacity of the compressor and a hot valve cover over the broken valve. The loss of capacity is caused by the valve not functioning and the heat over the cover is caused by the air going back and forth over the valve.
2. Leaky valve. Higher pressure in the stage or intercooler before the leaky valve. There may be a slight loss in capacity.
3. Four hundred °F is the maximum temperature in any stage. This is the lube oil flash point. The oil will form carbon deposits on the valves and possibly dieselize the compressor if the cylinder temperature nears, 400 °F.
4. Leaky first stage suction valves. Pulsating air at the compressor inlet filter.

Starting Air Systems

Direct-air starting systems are used on large diesel engines. Ship service generators are usually the smallest engines that use this system. The cylinder heads must have room for the air starting valve. Air is admitted a little past TDC on the power stroke. This is because all valves or ports are closed and the air can push on the piston, rotating the engine it up to starting speed in the correct direction.

A minimum of five cylinders are required to have starting valves in a four-stroke cycle diesel engine to start it in any position. A minimum of three cylinders are required in a two-stroke cycle diesel engine to start the engine in any crankshaft position. Usually, only one bank of a V-engine has starting air valves.

Most diesel engines use a pilot-operated system to open the balanced air start check valves located in the heads of the engine. The pilot valves usually are located in the camshaft area.

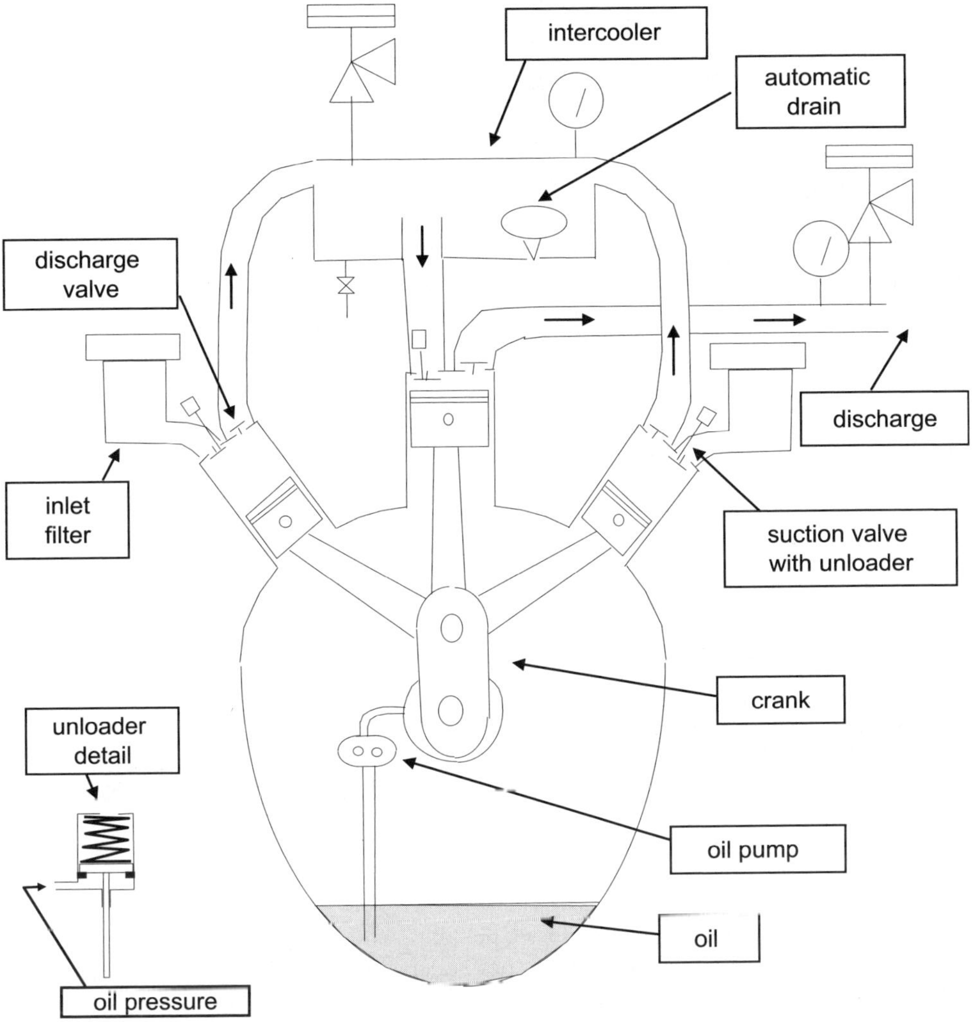

Fig. 9.3. Main starting air compressor W configuration

The pilot valves are arranged in an inline configuration for reversible four-stroke engines. They are arranged in a radial configuration for nonreversible four-stroke engines and slow-speed engines.

The balanced air start check valve has a balance piston in the center of the stem to balance the main starting air pressure. Thus, the balanced air check valve will not open, due to main starting air pressure alone. The pilot valve has to admit pilot air above the balance piston to open the valve. If the cylinder is firing, the firing pressure forces the valve closed, acting as a check valve. The balanced air start check valve has a light spring to keep it closed when it has no pilot air or the remote starting valve is closed.

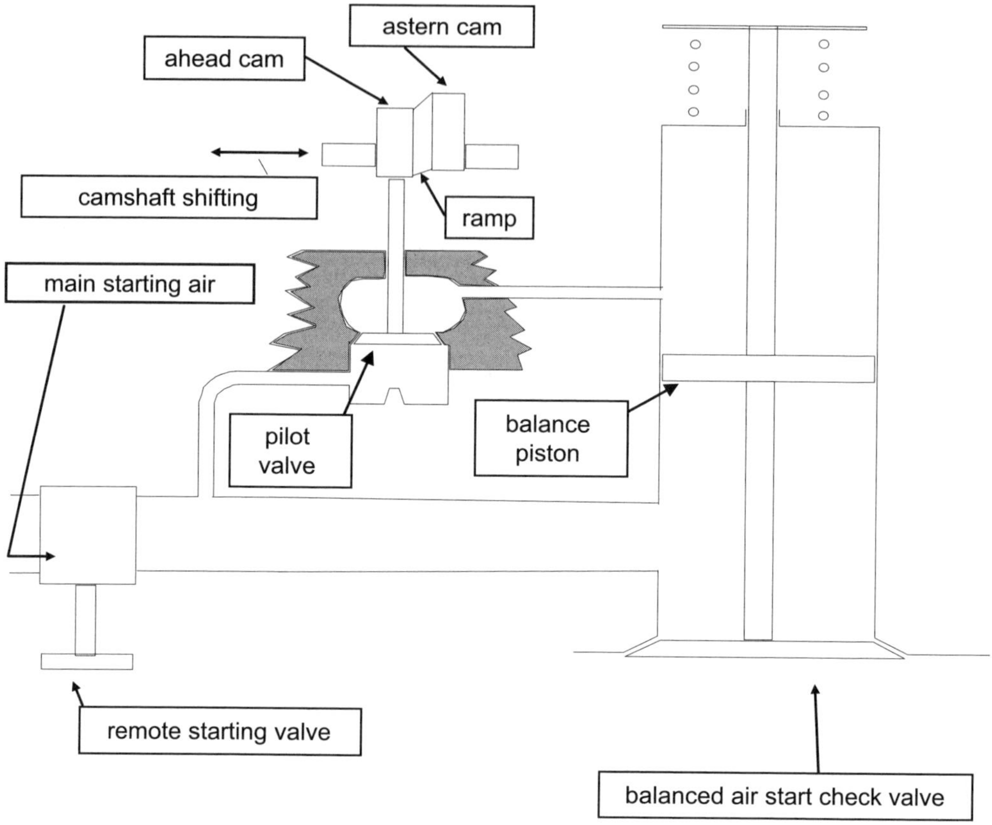

Fig. 9.4. Pilot air starting system.

When the remote starting valve is closed, the pilot valves drop down and away from the cam. Thus, no parts rub when the engine is not being started, which is most of the time.

Two cams are associated with the pilot valve. One is for ahead and one is for astern. The above system would be used on a four-stroke cycle diesel engine because the timing must be completely changed to go from ahead to astern.

The normal sequence in a four-stroke diesel engine is: intake-compression-power-exhaust-intake-compression. The reverse is: compression-intake-exhaust-power-compression-intake. The second sequence does not make sense. The timing of the engine camshaft must be set up with two cams for each function—intake, exhaust, fuel, and starting. The timing sequence must be correct for either direction, requiring a complete set of cams for each direction. Changing the cams requires shifting of the camshaft axially.

The second sequence works somewhat, which is the engine running ahead with the reverse cams. The engine takes intake air from the exhaust pipe and blows exhaust out the air intake.

This is the sequence in a two-stroke cycle diesel engine: compression-power-compression-power-compression. This sequence works backward or forward. To go from ahead

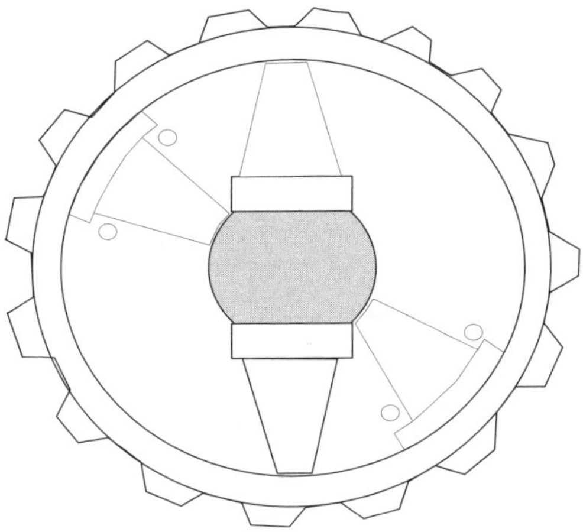

Fig. 9.5. Two-stroke direct reversing engine camshaft drive

to astern the camshaft has to be rotated from a little before TDC to a little after TDC. For engines using this technique, the cams must be symmetrical.

The gear is driven by the crankshaft through timing gears or roller chain. The central hub is attached to the camshaft. If the gear turns clockwise, the projections attached to the gears push the central hub along. If we want to reverse the direction, we have to reverse the timing. The above apparatus (fig. 9.5) is sandwiched between two plates and filled with oil contained by the appropriate seals. If oil is pumped in the holes located in the smaller volumes, and drained from the holes located in the larger volumes, the central hub rotates to the reverse timing. When the engine rotated, it pushes the hub in the counterclockwise direction. The apparatus is a sloppy camshaft drive that can be rotated with oil pressure within the stops provided.

A distributor can replace individual pilot valves in a two-stroke cycle engine that can be directly reversed or a four-stroke cycle engine that cannot be directly reversed.

The distributor, shown in fig. 9.6., is set-up for a four-cylinder engine, but any number of cylinders can be accommodated by having more holes and outlet pipes in the stationary disk.

The direct-reversing engine must stop, change the timing for the direction desired, and then start-up in the desired direction. To do this, the following sequence of events must take place:

1. Fuel is shut off to the injectors.
2. The timing of the engine is changed.
3. The main starting air valve is activated, putting air to the engine.
4. After the engine is rotating in the correct direction, fuel is turned on to the injectors.

Interlocks must be incorporated into the starting and reversing controls to ensure this happens. These controls are essentially an air-operated analog computer. The following interlocks usually are incorporated:

1. Cam position.

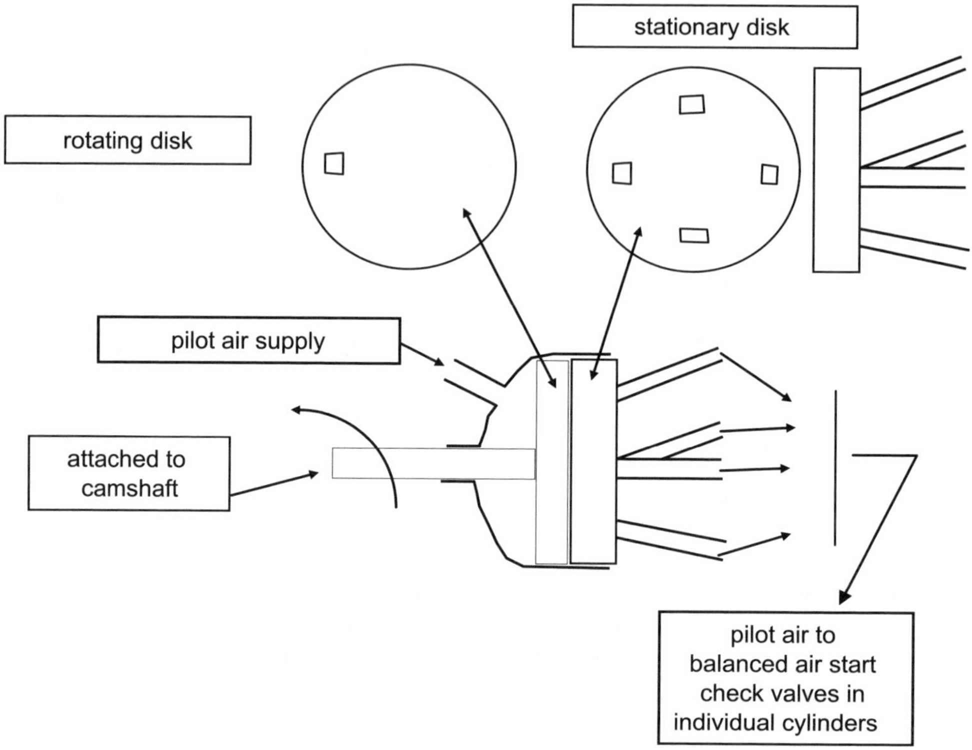

Fig. 9.6. Pilot air distributor

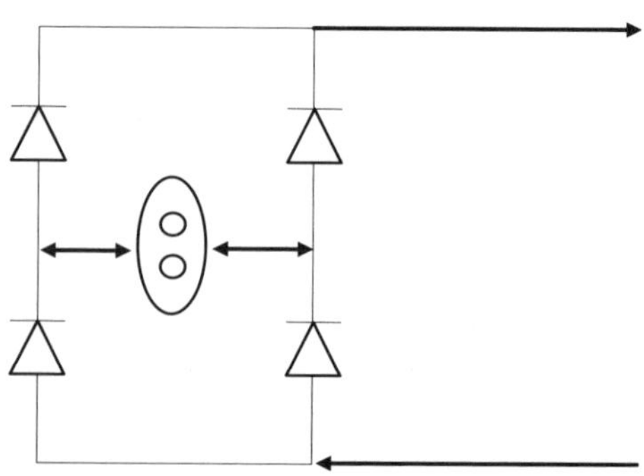

Fig. 9.7. Check valve arrangement for reversible rotary pumps

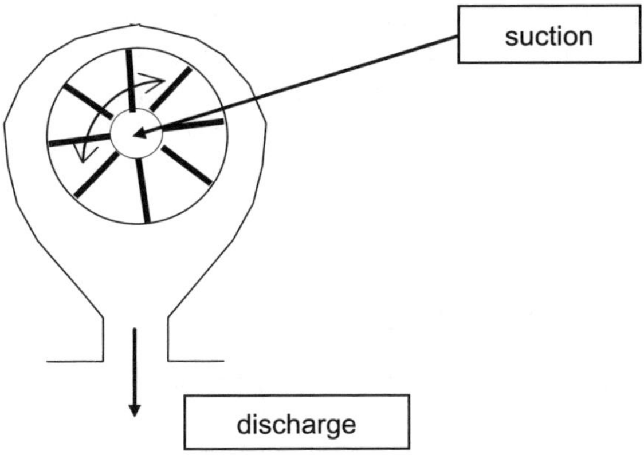

Fig. 9.8. Straight-vaned reversible centrifugal pump

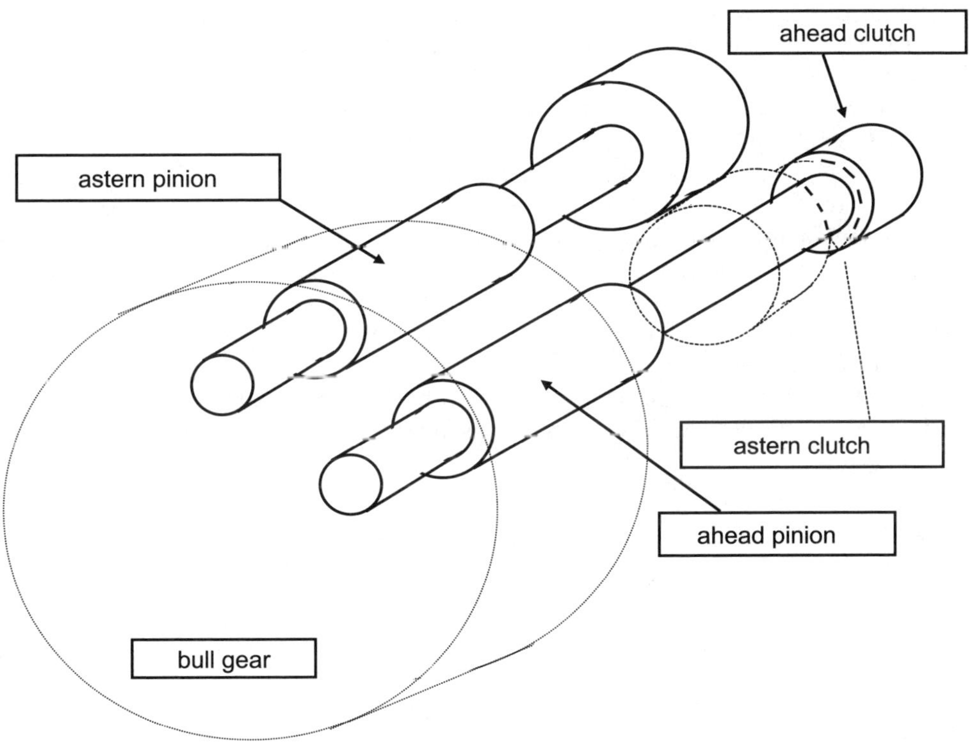

Fig. 9.9. Diesel reduction and reverse gear

2. Direction of rotation. The cam position must be verified before air is put on the engine. If the engine is still rotating in the original direction, the starting air with the new timing acts as a brake, stopping the engine, then starts rotating it in the correct direction. When the correct direction is verified, fuel is admitted to the cylinders starting the engine in the desired direction.
3. Lubricating oil pressure.
4. Cooling water flow or pressure.
5. Others, such as critical temperatures and turning gear engaged.
6. These interlocks also are incorporated into the system to prevent the operators from running the engine to destruction.

Attached pumps also are another item that must be taken into consideration when reversing an engine.

1. Reciprocating pumps are not affected because the suction and discharge valves within the pump will prevent the suction and discharge from changing places.
2. Rotary pumps, such as gear pumps for lubricating oil, will swap suction and discharge if the rotation is reversed. Four check valves set up like a full-wave rectifier will solve this problem.
3. Attached centrifugal pumps must have straight-vaned impellers and symmetrical volutes. These are used for cooling water.

Naturally electric motor driven pumps have no problems.

For high-speed and smaller, medium-speed engines, a reduction gear set with a reverse gear can be employed. Thus, the engine always rotate in the same direction.

The astern clutch is attached to the intermediate gear and rotates on the ahead pinion shaft. It is not attached to the ahead pinion shaft. Thus, if the ahead clutch is engaged, the ahead pinion rotates the bull gear and the astern elements freewheels in the opposite direction. If the astern clutch is engaged, the astern pinion rotates the bull gear in the opposite direction. This is because of the intermediate gear and the ahead elements freewheels in the opposite direction. If both clutches are engaged simultaneously, the gear lockups and stalls the engine before something breaks.

For slow- and medium-speed engines a controllable, reversible pitch propeller may be used. Slow- and medium-speed engines always rotate in the same direction using a CRPP.

Study Questions

1. Small starting motors use what device to engage the engine?
   A. electric clutch
   B. bendix
   C. solenoid
   D. cone clutch
2. Air is admitted to a direct air-start engine on which stroke?
   A. intake
   B. compression
   C. power
   D. exhaust
3. Large main starting air compressors are cooled by:
   A. oil
   B. air
   C. water
   D. Freon
4. A compressor unloader prevents the compressor from working by holding the:
   A. discharge valve open
   B. suction valve open
   C. electric contact open
   D. brake on
5. What opens a balanced air-start check valve?
   A. main starting air
   B. push rod
   C. spring
   D. pilot air
6. Which type of attached pump will cause problems in a direct-reversing engine?
   A. rotary
   B. gear pump with four check valves
   C. straight-vaned impeller
   D. reciprocating
7. If both clutches engage in a reverse gear the:
   A. gear will freewheel
   B. engine may stall
   C. governor will go to no fuel
   D. engine will over speed
8. What is not a device for storing power for a small-starting motor?
   A. capacitor
   B. battery
   C. accumulator
   D. air tank

9. What gas is used in an accumulator?
   A. air
   B. carbon dioxide
   C. argon
   D. nitrogen

10. This object represents a:
   A. differential piston
   B. castellated piston
   C. crosshead
   D. double-diameter piston

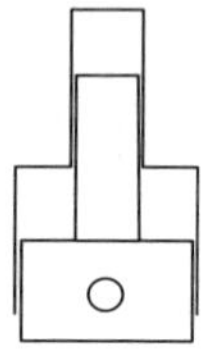

11. Four-stroke engines can be reversed by____________.
   A. shifting the camshaft
   B. rotating the camshaft
   C. A or B
   D. neither A nor B

12. In a direct reversing engine, starting air is admitted to the engine when:
   A. the engine rotation has stopped
   B. before the timing has been changed
   C. after the timing has been changed
   D. the engine rotation is in the desired direction

13. Medium-speed engines may use what type of reduction gears?
   A. single reduction
   B. double reduction
   C. A or B
   D. none of the above

14. Two-stroke engines are reversed by__________.
   A. shifting the camshaft
   B. rotating the camshaft
   C. A or B
   D. neither A or B

15. Two-stroke engines can be started from any position if they have____________ cylinders equipped with air start valves.
   A. three
   B. four
   C. five
   D. six

16. A distributor is used in the starting air system of what engine?
   A. nonreversing four-stroke engine
   B. direct reversing two-stroke engine
   C. A and B
   D. none of the above

17. In a direct-reversing engine fuel is admitted to the engine when:
    A. the engine rotation has stopped
    B. before the timing has been changed
    C. after the timing has been changed
    D. the engine rotation is in the desired direction

18. Four-stroke engines can be started from any position if they have____________ cylinders equipped with air start valves (minimum number required).
    A. three
    B. four
    C. five
    D. six

19. Pilot valves in an air start system engage the camshaft:
    A. during starting
    B. during shutdown
    C. when the jacking gear is engaged
    D. when running at a steady speed

Chapter 10
Governor Systems

Governors

Governor Characteristics:

1. Speed droop. This is the decrease in engine rpm from no load to full load expressed as a percent of the rated load rpm.
2. Isochronous governing. This is a governor that has no speed droop. The exact same speed is maintained regardless of load, as long as the prime mover is not overloaded.
3. Sensitivity. This is the change in rpm required to make the governor move the fuel rack, expressed as a percentage of the rated rpm.
4. Stability. This is the ability of the governor to maintain a constant speed without short term variations. In other words, no hunting.
5. Hunting. These are fluctuations in speed with no load change.
6. Promptness or response. Speed of action from no load to full load expressed in seconds.
7. Power or work capacity. The amount of force or power the governor can exert to move the fuel racks.

There are three types of governors: electronic, direct acting, and relay.

Electronic Governor

The electronic governor is a computer-based system using the following components:

- computer
- analog to digital converter
- digital to analog converter
- servomotor
- sensors

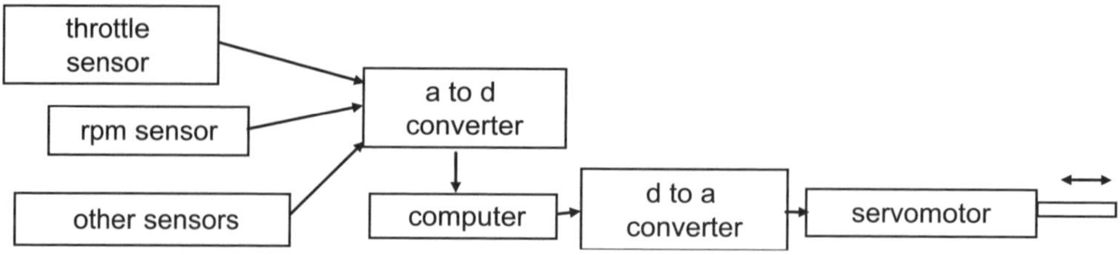

Fig. 10.1. Electronic governor

The rpm sensor, throttle sensor, and other sensors, such as intake manifold pressure, send analog signals to the A to D converter. The A to D converter digitizes the signals and sends them to the computer. The program in the computer analyzes these signals and compares them to the desired input from the throttle sensor. The computer sends the desired outcome to the D to A converter. This analog signal is then sent to the servomotor that controls the fuel racks.

This system is very flexible and governs isochronously. Many more inputs, such as maximum cylinder pressure or an ideal pressure volume diagram, can be input to the computer to control an electronic fuel injection system to run the engine more efficiently.

Direct-acting Governor

This type of governor uses spring-loaded flyweights to move the fuel rack and control engine speed.

The direct-acting governor depends on the centrifugal force of the flyweights to move the speeder rod that is attached through a suitable linkage to the fuel rack.

Centrifugal Force $F = ma_n \quad a_n = \omega^2 r = \frac{V_t^2}{r} \quad V_t = r\omega$

Spring Force $F = kx$

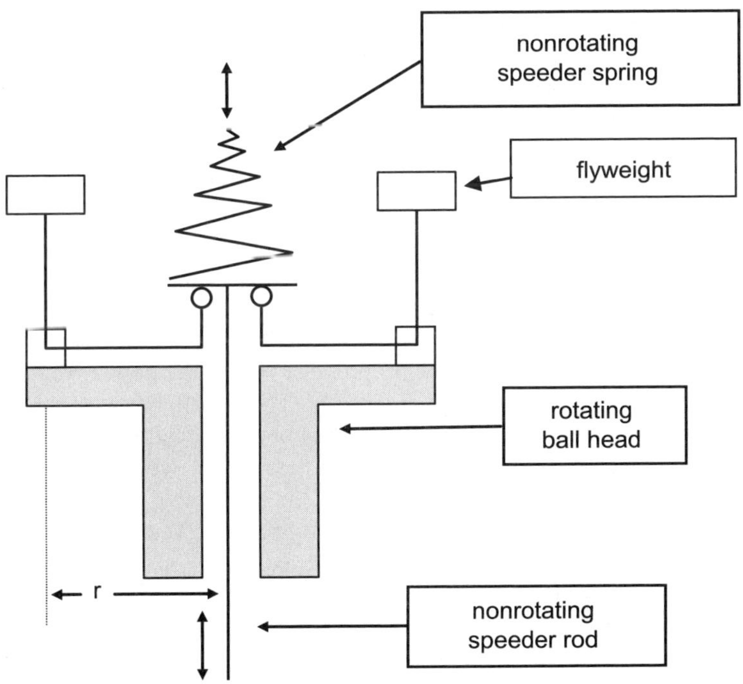

Fig. 10.2. Direct-acting governor

The centrifugal force is a function of the square of the rpm. The flyweights produce this force that is acting against the spring force. The spring force usually is a linear function for a cylindrical spring, creating a nonlinear force opposing a linear force. This does not produce the same movement of the speeder rod with the same change in rpm over the rpm range of the engine. Notice, in our governor we have a conical or trumpet-shaped spring. This is a nonlinear spring that matches the centrifugal force and produces the same movement of the speeder rod over the rpm range of the engine.

This direct-acting governor always has speed droop and cannot be made isochronous. This is because the only way to get at the exact same speed, the flyweights and the speeder rod attached to the fuel rack must be in the same exact position. If a load is put on the engine, the flyweights move in, lowering the speeder rod to put more fuel on the engine. The governor cannot go back to the original position because it would be running at the same rpm with more load at the same fuel setting.

This governor also has limited power since the centrifugal force of the flyweights and the spring are the only motivating forces. It also has poor sensitivity, since the centrifugal and spring forces are the only things moving the fuel rack and overcoming friction.

The advantage of this governor is simplicity. It is only used in small engines.

Relay Governor

The relay governor overcomes the disadvantages of the direct-acting governor by adding a hydraulic power circuit, thus, the name relay governor. The governor contains a gear pump and pressure regulator. A pilot valve moved by the speeder rod controls the oil pressure. The oil controlled by the pilot valve actuates a hydraulic piston that moves the fuel rack. We now have potential for unlimited power. Greater sensitivity is obtained, since the flyweights and spring can be designed for sensitivity and not power.

This governor also is isochronous because the flyweights can return to the original set position with the power piston and fuel rack in a new position, depending on the load.

The problem with the relay governor, shown in fig. 10.3., is that when the pilot valve sends oil to the power piston, it goes full stroke. When the pilot valve dumps the oil from the power piston, the reverse happens. This means that the governor is unstable and hunts.

Which direction of the power piston moves the fuel rack in the more fuel direction? If you put a higher load on the engine, it slows down. The flyweights move in toward the center. The pilot valve moves down. The oil pressure moves the power piston up, increasing the fuel. The opposite happens if you reduce load on the engine.

This problem of hunting can be cured by adding negative feedback into the system.

In fig. 10.4., a negative feedback bar has been inserted between the power piston and the pilot valve. It is negative feedback because as the pilot valve moves down, the power piston moves up and starts to move the pilot valve back up. We have now eliminated the hunting because the power piston cannot go full stroke, but stops itself through the negative feedback bar. The problem with this is that the governor always has speed droop. The governor is not isochronous, because the power piston has moved the pilot valve before the flyweights start to return to the original speed. The pilot valve blocks all oil flow before the original speed is reached.

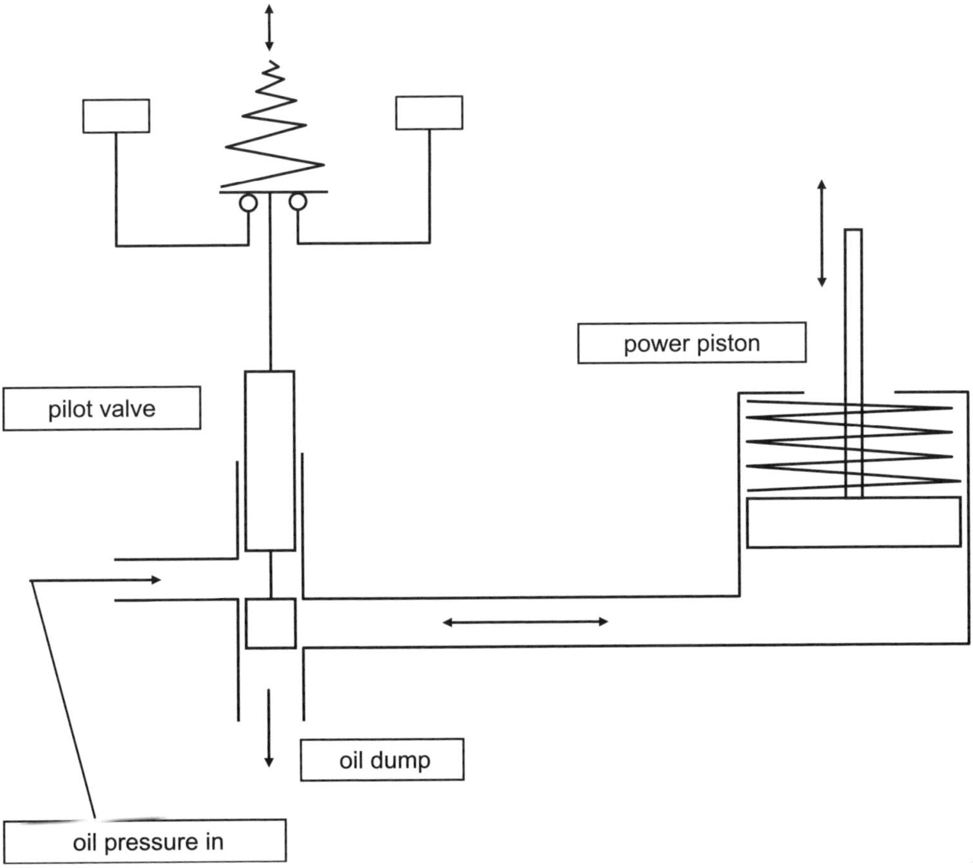

Fig. 10 3. Relay governor

Which direction of the power piston moves the fuel rack in the more fuel direction?

The problem of inherent speed droop can be solved by using a compensation circuit that essentially puts in temporary negative feedback.

If the needle valve in the small oil reservoir is wide open, then we have the first relay governor with no negative feedback and with the instability this brings. This is because the sending piston goes up and down without having any effect on the receiving piston.

If the needle valve is shut completely, we have the standard negative feedback relay governor with speed droop. This is because the receiving piston always follows the sending piston and provides negative feedback, just like the feedback bar connecting the power piston and the pilot valve.

If the needle valve is throttled, then negative feedback reaches the pilot valve, but the centering spring brings back the receiving piston to its original position after a short time delay. The flyweights have to be in their original position to stop the flow of oil to and from the power piston. The only way this happens is if we are back at the original speed with

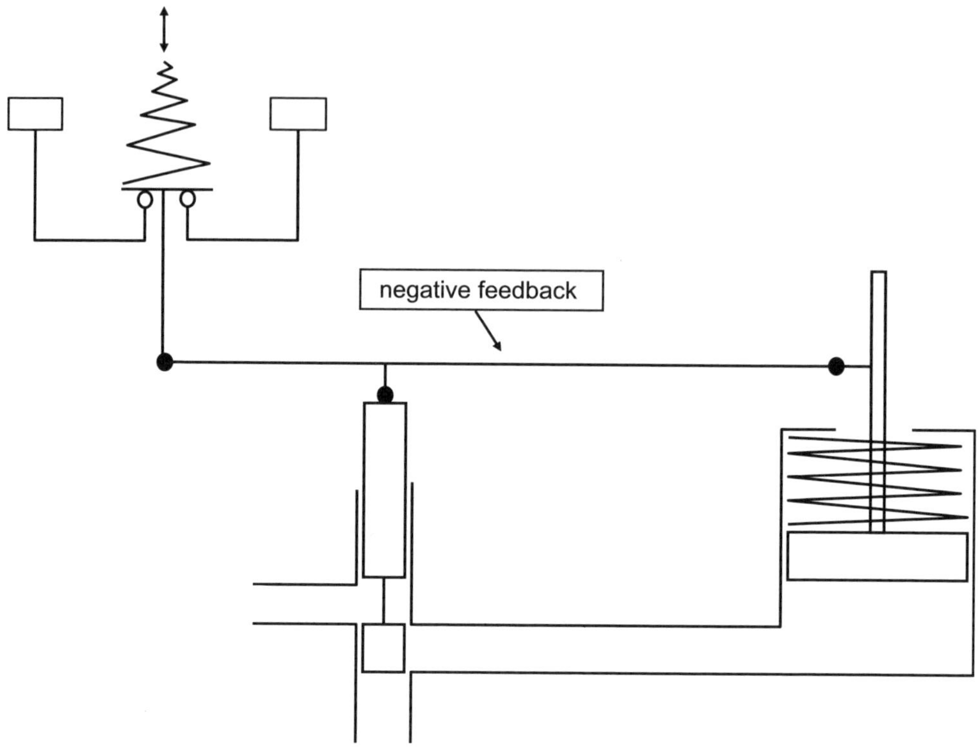

Fig. 10.4. Relay governor with negative feedback

a new load and a new position of the power piston and the sending piston. Therefore, we have an isochronous governor that is stable.

Which direction of the power piston moves the fuel rack in the more fuel direction?

Most governors have three control knobs and a lever on their face:

1. Speed. This compresses and decompresses the speeder spring to raise or lower the speed.
2. Speed droop. This controls the manual negative feedback bar as shown below. When the pivot point at the top of the speeder spring is aligned vertically with the end of the spring, no speed droop occurs. It is impossible to parallel two isochronous governed engines without load sensing circuits. They would end up driving each other because it would be impossible to dial in the exact same speeds. Thus, it is important to be able to dial in a speed droop by manual means for one of the governors.
3. Load limit. This is a movable stop that limits the power piston travel and the maximum amount of fuel to the engine supplied.
4. Compensation lever. This lever adjusts the needle valve in and out to adjust the amount of compensation.

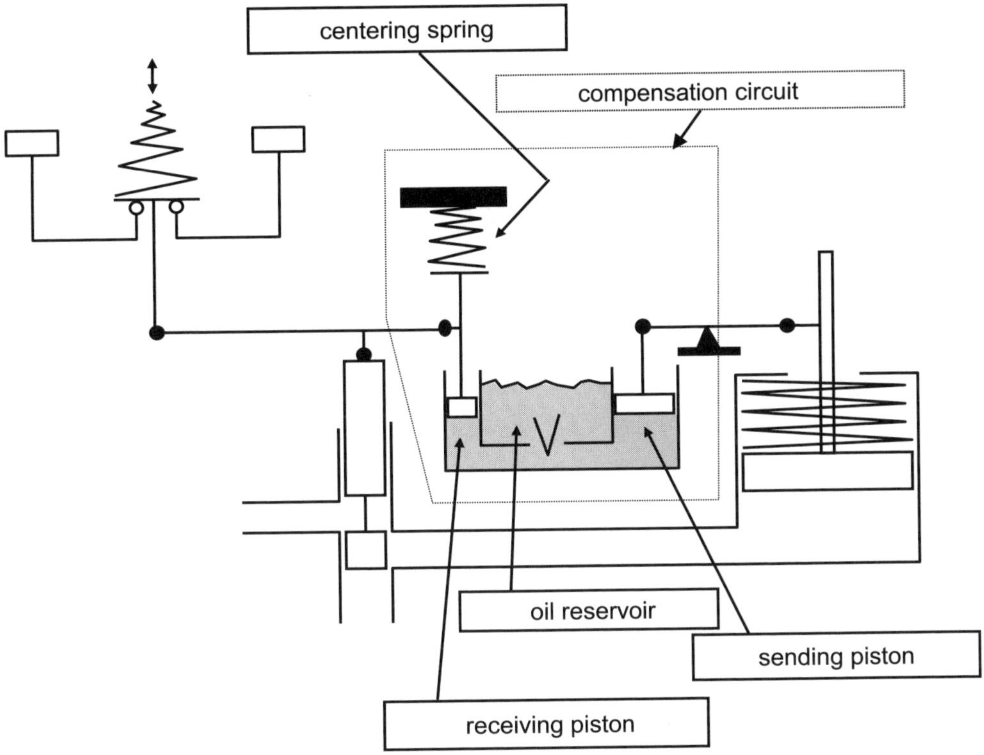

Fig. 10.5. Relay governor with compensation circuit

The generator with the lowest speed droop or shallowest slope takes the most load. Fig. 10.7 shows two AC generators with different speed droops dialed in on their governors. They are AC generators in parallel operation so they must have the same speed.

Fig. 10.8, a generator with speed droop and a generator with no speed droop are in parallel. In this way, part of the load can be done isochronously with no load sharing devices. Most modern plants have load-sharing circuits, so the whole load range would be governed isochronously.

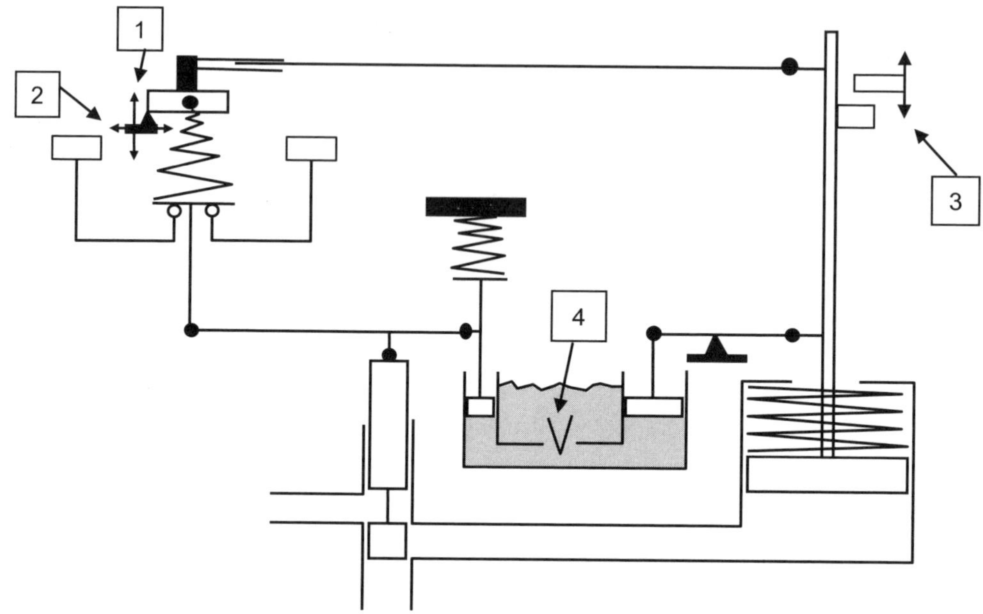

Fig. 10.6. Full feature relay governor

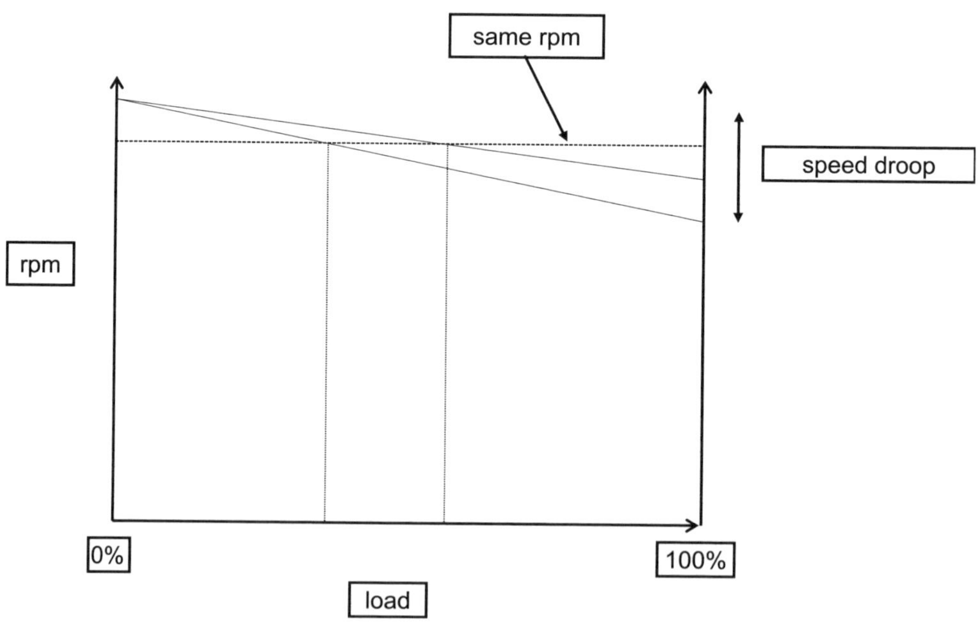

Fig. 10.7. Two generators at different speed droops

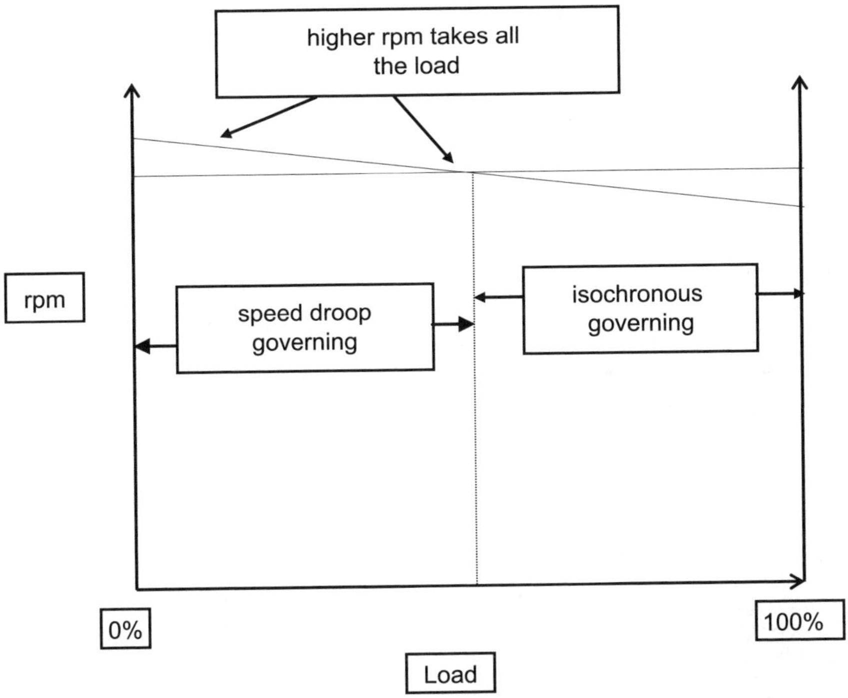

Fig. 10.8. Two generators, one with speed droop, one isochronous

Study Questions

1. Which of the following is not a basic type of governor?
   A. electronic
   B. direct acting
   C. indirect acting
   D. relay
2. A governor acting isochronously:
   A. can be direct acting
   B. can have speed droop
   C. cannot have speed droop
   D. does not have a power circuit
3. In diagram A, which device controls the oil to the power piston?
   A. 2
   B. 3
   C. 4
   D. 5

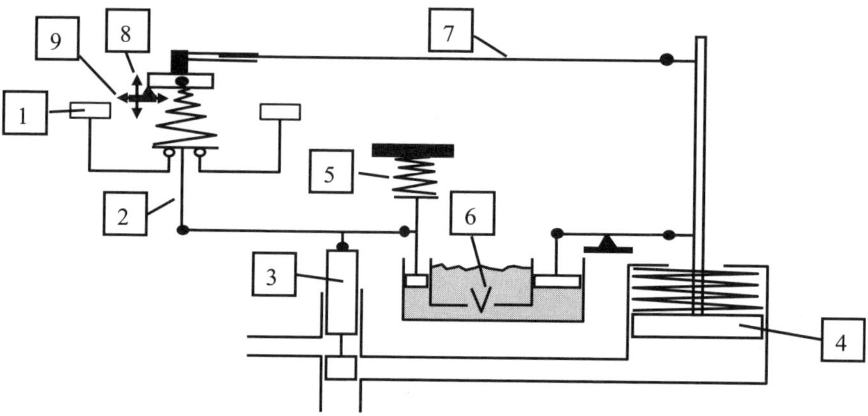

Diagram A

4. In diagram A, which device is adjusted to eliminate hunting?
   A. 6
   B. 7
   C. 8
   D. 9
5. The device represented in diagram A is a:
   A. relay governor
   B. direct acting governor
   C. electronic governor
   D. hydraulic pump

6. In diagram A, which device is the power piston?
   A. 2
   B. 3
   C. 4
   D. 5
7. In diagram A, which device is adjusted to add mechanical negative feedback?
   A. 6
   B. 7
   C. 8
   D. 9
8. In diagram A, which device is adjusted to change the governing speed?
   A. 6
   B. 7
   C. 8
   D. 9
9. Sensitivity is the change in _____________ required to make the governor move the fuel rack.
   A. time
   B. load
   C. input
   D. rpm
10. The speeder spring on most governors is:
    A. linear
    B. nonlinear
    C. A or B
    D. none of the above
11. When two AC generators with governors that have two different speed droops are paralleled:
    A. the generator with the higher speed droop will run at a higher rpm
    B. the generator with the lower speed droop will run at a lower rpm
    C. the generator with the lower speed droop will run at a higher rpm
    D. the generators will run at the same rpm
12. Which governor has a hydraulic circuit?
    A. electronic
    B. direct acting
    C. relay
    D. indirect acting

Chapter 11
Fuel Injection Systems

Fuel Injection System Requirements

Fuel injection systems of diesel engines have the flowing requirements:

1. Accurate metering to each cylinder. The exact same amount of fuel must be metered to each cylinder of the engine or some cylinders will be overloaded and some will be underloaded.
2. Proper timing. The fuel must be delivered to the cylinder at the proper time before TDC on the compression stroke. Each cylinder must have the same timing. If this does not occur, the combustion pressures between the cylinders will vary, even if the same amount of fuel is delivered to each cylinder. The PV diagram for each cylinder should be the same.
3. Suitable rate of injection. The rate of injection dictates the pressure rise in the cylinder. Too high a rate sets up high stresses and vibration in the connecting rod and the crankshaft. The rate of injection also can be used to make the pressure volume diagram appear as either an Otto cycle or Diesel cycle. Thus, the engine can be tuned for the greater efficiency of the Otto cycle or the greater power of the Diesel cycle.
4. Proper atomization. The fuel must be broken down into small droplets to start combustion. This is accomplished by forcing the oil through small holes at high pressures.
5. Proper penetration. The fuel oil must travel throughout the combustion chamber to reach the oxygen. However, the oil must not travel so far as to hit the walls of the combustion chamber and cylinder liner. Here the fuel would cool off and wash away the lubricating oil film on the liner.

The same variables affect both atomization and penetration. The smaller the droplet, the more surface area the drops have for heating and combustion. However, this means a smaller mass and less penetration. Very high injection pressures can create both small droplets and high penetration. These parameters must be customized for each combustion chamber geometry and fuel injection system.

Types of Injection Systems

1. Air Blast. The fuel pump in this system metered low-pressure fuel into a fuel valve. The fuel valve had a high-pressure air connection. The high-pressure air was controlled by cam action timed to the crankshaft. Just before TDC, the high-pressure air would blast the fuel into the combustion chamber with a small amount of air. This required a large, high-pressure air compressor that used a lot of energy. This system is not used today.

The fuel injection systems used today are considered solid injection systems. We are not injecting solids, but an incompressible liquid.

2. Common Rail. This system uses a single, high-pressure pump and pressure regulator to produce the high-pressure needed for injection. This high pressure is fed to a pipe that runs along the engine near the injectors. This pipe is called a common rail. The injectors are fed high-pressure fuel from this common rail. They originally were activated by cam action timed to the crankshaft to inject the fuel just before TDC on the compression stroke. The problem with this system is that it is hard to control small amounts of fuel. The injectors would allow large quantities of fuel in the cylinder if they leaked. These problems were mainly due to the mechanical cam-operated injectors. This system is no longer used, but a modern version is available. Instead of using mechanically operated injectors, the new system uses solenoid-operated injectors controlled by a computer.
The computer-controlled common rail system can be precise and operate at very high speeds. It can be used on high-speed engines, slower main propulsion engines, and for pilot injection. Once mechanical injectors open, they can only close at the end of injection. This means the first fuel sprayed into the combustion chamber rises up to the autoignition temperature. This ignition delay allows more fuel to spray into the combustion chamber without igniting. Finally, the fuel charge goes off with an explosion. The solenoid-operated injectors are fast enough so that a small quantity of fuel is first injected and starts to burn. The main charge then can be injected into the burning fuel, promoting better combustion.
The future common rail systems will be able to generate pressures of 1,300-1,600 bar. Because the high-pressure pump is not directly dependent on a cam-operated piston pump, this pressure is available at idle speeds. This promotes better atomization and

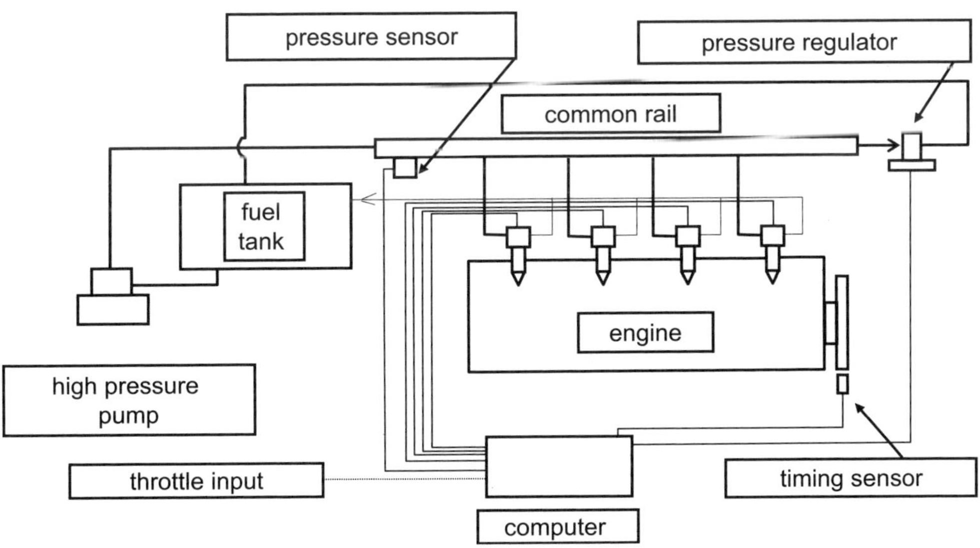

Fig. 11.1. Modern computer-controlled common rail system

more complete combustion throughout the speed range of the engine. This way produces less smoke, even at idle. The system can be turned, so that emissions of nitrogen oxide (NOx) and particulate matter are reduced. This aids the diesel engine in emission standards.

3. Jerk Pumps. The basic principle of operation of these high-pressure fuel pumps is to close all outlets of a piston pump, except for the outlet to the differential hydraulic injector. This injector acts as a safety valve. The injector pops when its set pressure is reached and sprays fuel into the combustion chamber. The high-pressure piston pump operates from the camshaft of the engine. The pump and the injector have no packing. The high pressures are contained by finely mated surfaces and close tolerances. Most manufacturers use a type of port and helix pump to control the amount of fuel output. Sulzer is the only manufacturer who uses lever-operated suction and spill valves to control fuel output on their slow-speed engines.

The common rail system probably will be the most common system in the future. However, fuel injection system manufacturers have designed computer-controlled systems of jerk pumps by using a solenoid to open and close the ports of the pump.

The high-pressure fuel injection pump diagrammatical representation (fig. 11.2) is of the Bosch-type. Most of the modern fuel injection pumps are of this type. The only differences between the pumps are the way the port and helix are arranged.

The jerk pump shown in fig. 11.2. works in the following manner:

1. The ports are surrounded by fuel supplied by the fuel oil service system at low-pressure.
2. As the cam pushes the plunger upward, the ports close and the fuel has no place to go but through the discharge valve to the injector. This is the beginning of injection.
3. Fuel no longer goes to the injector when the helix uncovers one of the ports. This is the end of injection.
4. The cam continues to rotate and a return spring (not shown) keeps the cam follower in contact with the cam as the follower drops down to the base circle and the plunger drops down to its lowest position. This system does not have lash, as that occurs in the valve train components.
5. As the top of the plunger drops past the ports, low-pressure fuel enters the barrel and the process starts over again.
6. The physical stroke always is the same and it is the same as the cam height. The effective stroke is changed by rotating the toothed control sleeve with the toothed rack (not shown). Since the plunger tangs engage the control sleeve slots, the plunger also rotates, changing the position of the helix in relationship to the ports in the barrel. This changes the effective stroke and thus, the amount of fuel going to the injector.
7. The plunger above has a constant beginning for injection. This would work for a constant-speed engine, such as a generator. A variable-speed engine, such as a main propulsion engine, would need a variable beginning of injection to advance the timing with higher speed. This is shown in fig. 11.3.

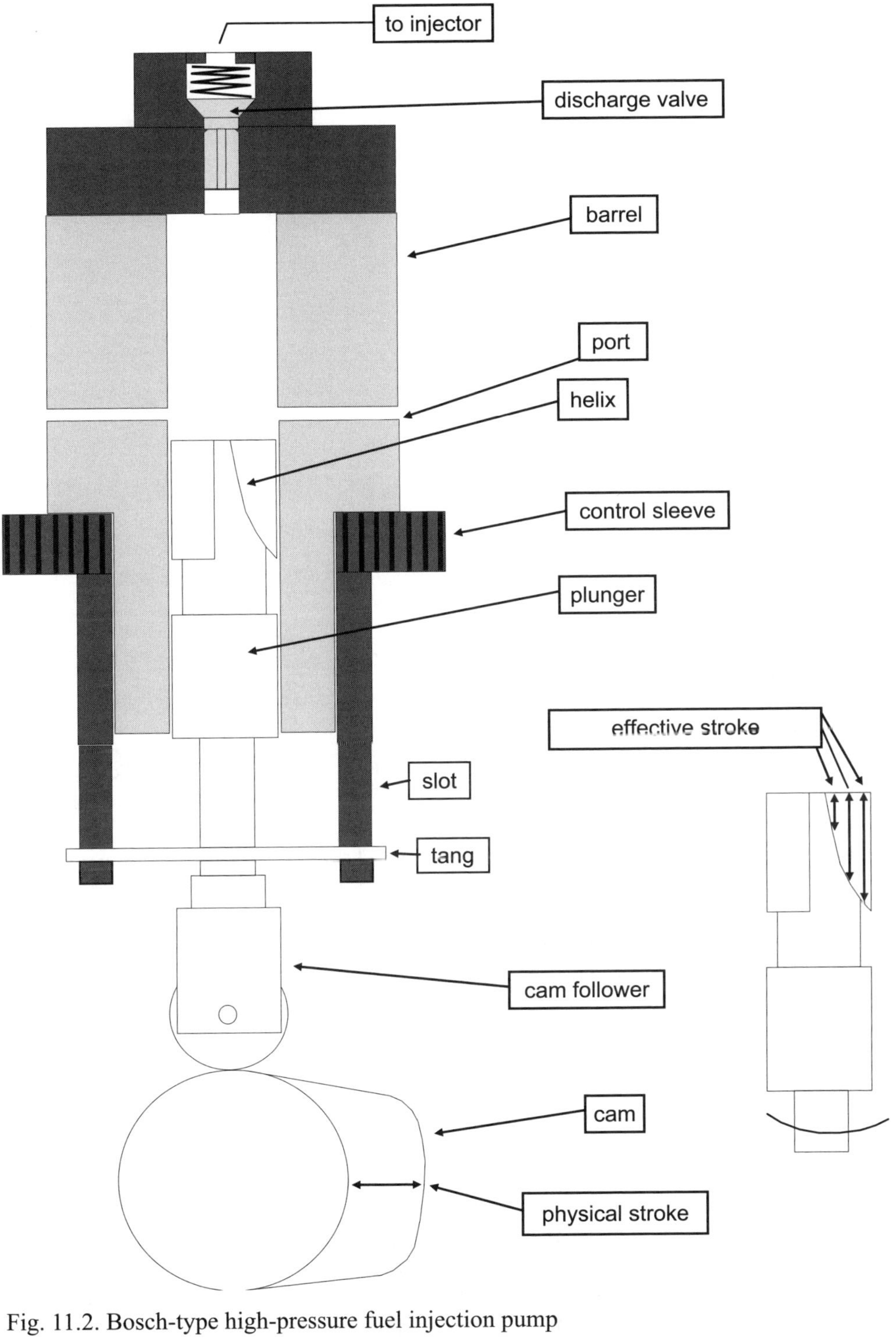

Fig. 11.2. Bosch-type high-pressure fuel injection pump

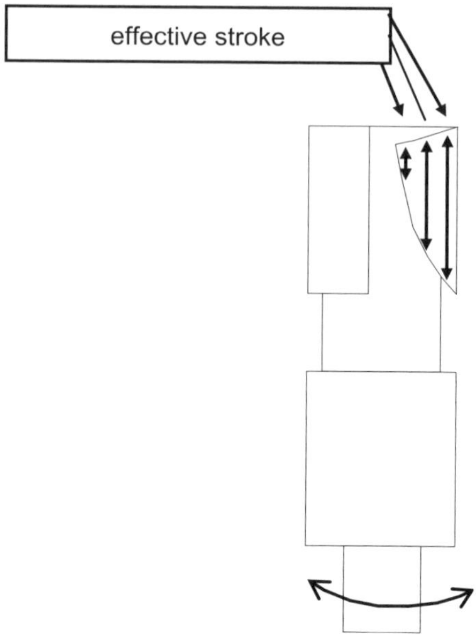

Fig. 11.3. Injection pump effective stroke

Discharge Valve Details

The discharge valve in medium- and high-speed engines is a special check valve with a retraction piston. The check valve part keeps the injection line pressurized. The retraction piston increases the volume above the valve at the end of injection. In medium- and high-speed engines, the high-pressure fuel pump and the injector are operating at such a rate that a shockwave is produced at the end of injection. This is when the spill port is uncovered by the helix. If the volume above the discharge valve is increased the right amount, the pressure in the line to the injector lowered and the shock wave is unable to reopen the injector. The retraction piston does not eliminate the shock wave, but eliminates the effect of reopening the injector and allowing fuel to enter the combustion chamber when it is undesirable.

The retraction piston volume has to be designed with the length of the line to the injector taken into account. That is why the injection lines on the same engine are all the same length. If the retraction piston is too small, the shock wave can reopen the injector because the pressure above the discharge valve will not have dropped enough. If the retraction piston is too large, the pressure drops too much, causing cavitation that can result in damaging parts of the system, especially the needle valve in the injector.

The discharge valve actually is closed in the position shown in fig. 11.4 because the retraction piston has entered the bore of the valve body. The spring will push the valve disk down until the slope seats contact each other. In doing this, the volume above the valve is increased. For this device to work, there has to be very close tolerances between the

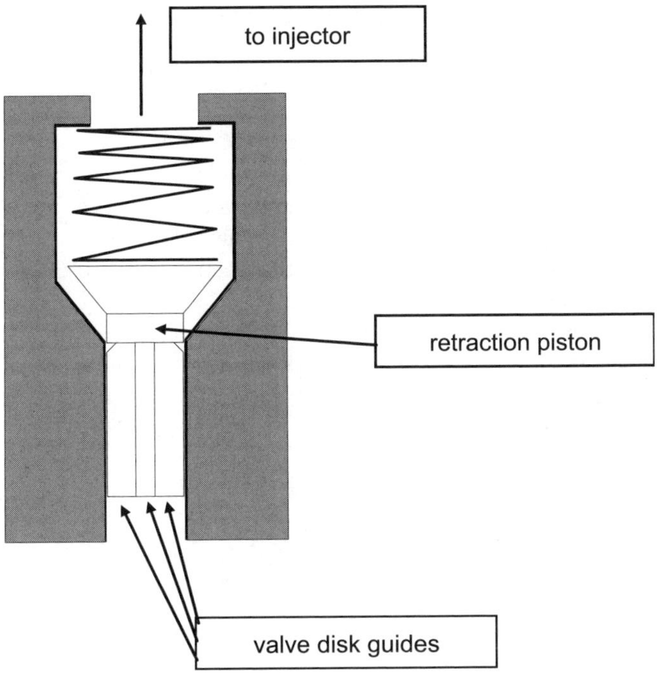

Fig. 11.4. Injection pump discharge valve

mating surfaces. This includes the small guides located under the retraction piston. Thus, everything about the high-pressure pump manufacturing is close tolerance, high stress and expensive.

Fuel Cam Design

The following must be taken into account when designing the cam:

1. The displacement of the cam follower must be continuous and even to keep the acceleration low. High acceleration produces heavy loads on the operating gear and scuff the cam and follower surfaces. Theoretically, the contact area between the cam and cam follower is a line with no width. This would lead to very high stresses because essentially there is no area between the mating surfaces. The surfaces, actually deform so that we do have an area for the force to act on and produce an allowable stress. The Hertz equations allow one to calculate the stresses involved. The limiting stress used for design is the fatigue limit. This is because of the cyclic nature of the load.
2. The ports of the high-pressure pump must be closed in an area of constant velocity. This will give the fuel a constant rate of injection and uniform atomization.
3. The plunger and cam follower must be decelerated so that zero velocity occurs at the top of the cam. The return spring must be designed to keep the cam follower in contact

with the cam during this period, and return the follower to the base circle. No bounce with the associated high-impact loads can be tolerated.

4. The back side of the cam must be a mirror image of the front side in direct-reversing two-stroke cycle engines.
5. In nonreversing two-stroke and four-stroke cycle engines, the trailing side of the cam can be designed for minimum loading of the cam surfaces and energy balancing throughout the camshaft. Energy is coming out of the camshaft as the plunger goes up and energy is returning to the camshaft from the return spring as the plunger goes down. This energy must be balanced so that resonant frequencies are not setup in the camshaft and the drive mechanism from the crankshaft.

   The shape of the cam can be generated from the above requirements, knowing that you can calculate the velocity from integrating the acceleration and the displacement from integrating the velocity. You also can go in the opposite direction by differentiation.

   To begin, we know that we need a constant velocity of the plunger when the ports are closed.

Fig. 11.5 represents the acceleration, velocity, and lift of the cam from the base circle to the nose or top of the cam.

Note the following:

1. The velocity is constant, the displacement or lift has a constant slope, and the acceleration is zero when the ports are closed.
2. The acceleration is maximum above the point of inflection in the velocity curve.
3. The area under the acceleration curve equals the value of the velocity curve. The area under the velocity curve equals the value of the displacement curve.
4. The value of the velocity curve is equal to the slope of the displacement curve. This is why we have a constant slope in the displacement curve, where we want the velocity constant. The value of the acceleration curve is equal to the slope of the velocity curve.
5. The result of integration is a higher order curve and the result of differentiation is a lower order curve.

These rules apply to all curves related by integration and differentiation.

After the above curves are calculated, the acceleration is checked with the Hertz equations to see if the surfaces will fail. The plunger velocity may have to be adjusted to keep the acceleration to acceptable levels. If the velocity of the plunger is decreased, the plunger diameter can be increased to keep the amount of fuel pumped the same.

This procedure can be used to design other cams, such as intake, exhaust, and air start, but with different constraints other than the constant velocity constraint of the fuel cam.

Another consideration for the fuel cam follower is that even though tolerances between the plunger and the barrel are close, some fuel will leak down the barrel. This is needed for lubrication, but on large engines this oil will mix with the lubricating oil of the engine. The geometry of the area under the high-pressure fuel pump is setup to catch this leakage and drain it to a waste oil tank.

Large engines usually have one high-pressure fuel pump per cylinder. Smaller engines have the pumps located in one unit with equal length injection lines going to each cylinder.

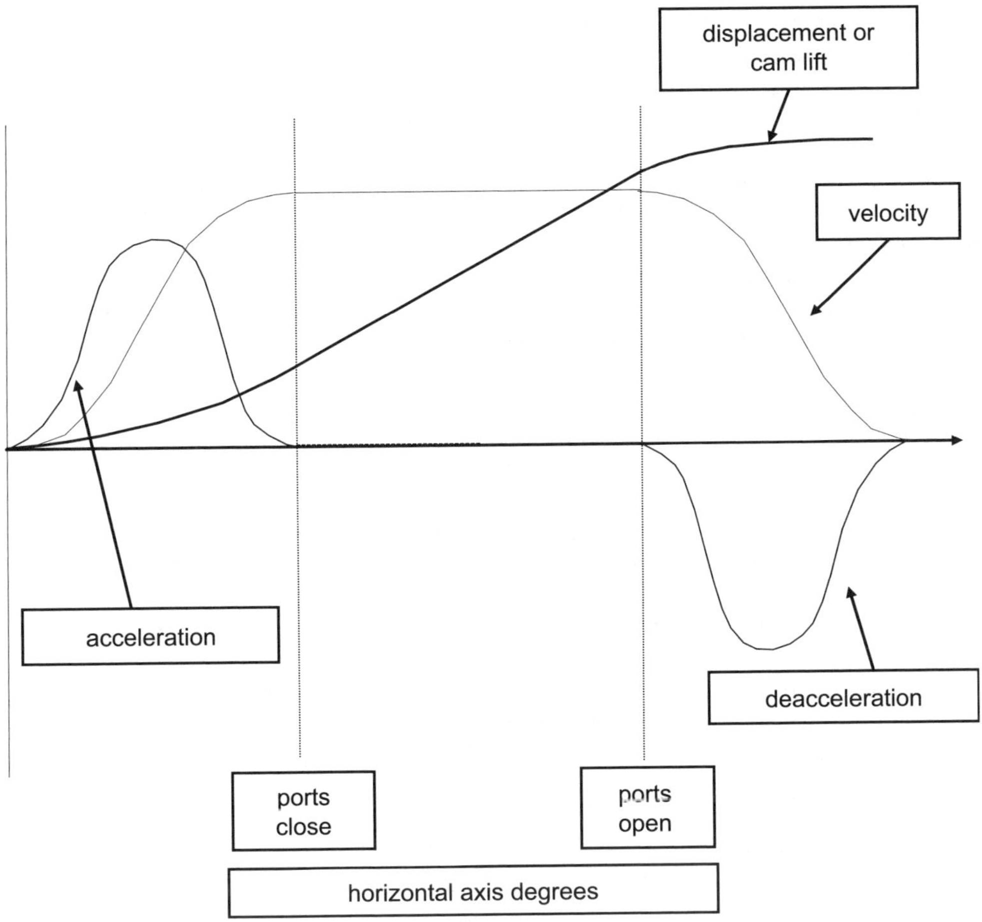

Fig. 11.5. Fuel cam design parameters

Camshaft Design

The load on the cam surfaces is cyclic in nature because the follower is rolling or sliding over the surface. This means that a different part of the surface is loaded as the cam rotates, cyclically loading every part of the surface. The cam surface must be designed for fatigue. This is best done by surface hardening. The cam is forged, machined, surface hardened, and ground to a smooth finish. The follower surface is manufactured the same way. The follower on most heavy-duty engines is the roller type.

The camshaft is driven by the crankshaft in a direct relationship to time the engine. This can be done by gears, roller chain or shafts, and bevel gears. Smaller engines may use a fiberglass toothed belt or link belt.

The camshaft usually is made of a one-piece forging. This is for small engines and some fairly large medium-speed engines. Larger engines use a built-up camshaft. The shaft is made up of flanged sections with the individual cams attached to the shaft. The cams can be split along the base circle and bolted to the shaft. Keyways, small gear teeth or another type of serration may be used. However, any of these devices can setup stress risers in the camshaft or cam. These can be eliminated by using an interference fit.

Fig. 11.6 shows how an interference fit can work without using any stress riser inducing serrations. The taper section is exaggerated greatly. The idea is to supply high-pressure oil to the cam and stretch the internal diameter so that it can slide onto the greater diameter machined in the camshaft. The oil pressure is then released when the cam is in the proper position. The cam is now attached to the camshaft with an interference fit.

Timing the high-pressure fuel pump to the engine is done by adjusting the ports and plunger distance. This can be done two ways with a port and helix pump:

1. Shims can be added under the pump body to raise the ports and retard the timing. Taking out shims will advance the timing.
2. A jack screw can be incorporated into the cam follower. This can be used to lower the plunger to retard timing or raise the plunger to advance timing. For a four-stroke cycle engine, the position of the plunger is determined at the start of injection. This would be so many degrees before TDC on the compression stroke. If the position of the plunger is determined just before TDC on the exhaust stroke by mistake, the adjustment of the jack screw will result in the plunger being too high, which will destroy the fuel pump and cam follower. This would not happen in a two-stroke engine because it has only one TDC in the cycle.

The rack setting is used to make sure each cylinder receives the same amount of fuel. The fuel rack has index markings. The pumps are matched so the index markings can be used to balance the engine initially. Indicator diagrams can be taken to fine-tune the fuel

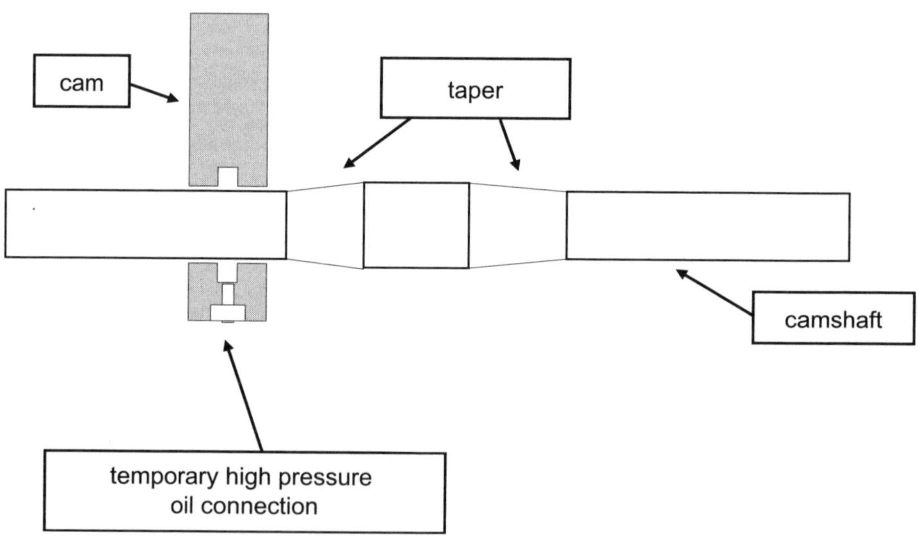

Fig. 11.6. Interference fit cam attachment

rack settings. However, if the readings of the index marks on the racks differ greatly, the pump probably is worn and should be replaced.

The Sulzer slow-speed engine, high-pressure fuel pump is operated by a levered suction valve to start injection and a levered spill valve to end injection. The levers are operated by a plunger with no helix. These valves are setup with dial indicators in relationship to the plunger and the plunger is set with a dial indicator to the flywheel. This takes care of both balancing and timing.

The Sulzer high-pressure fuel pump, as seen in fig. 11.7, functions as follows:

1. As the plunger rises, the suction valve closes, starting injection.
2. As the plunger keeps rising, the spill valve opens ending injection.

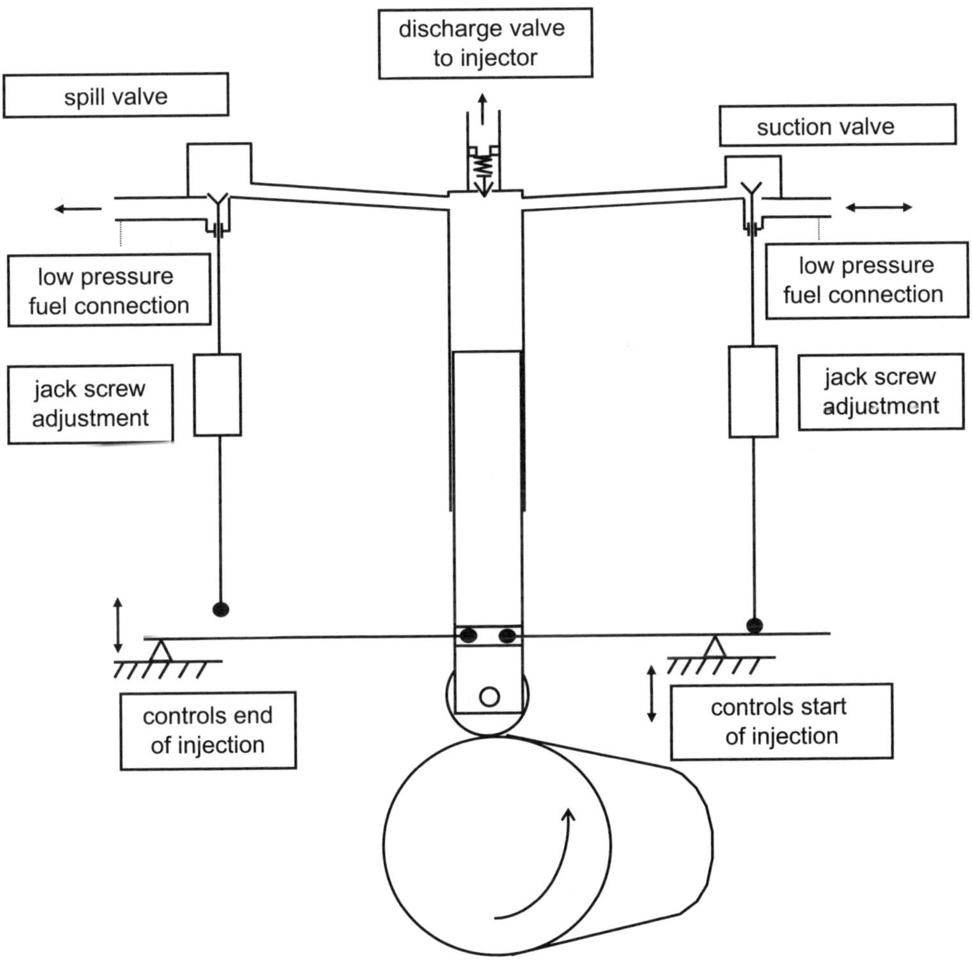

Fig. 11.7.Suzer fuel injection pump

The beginning and ending of injection can be adjusted by the length of the valve push rods.

As the engine is running, the governor will control the end of injection by moving the spill valve pivot point up or down. This is for older engines. Newer engines have a linkage between the pivot points of the suction valve and the spill valve to control the length of injection and advance the start of injection with speed. The faster the engine is running, the more advanced the timing needs to be.

We saw this same feature in the port and helix pump. However, we could not adjust this pump as easily as adjusting the linkage in the Sulzer slow-speed pump. This is because the advancement of the timing with speed was accomplished by machining of an upper helix. It is desirable to adjust the beginning and end of injection depending on the type of fuel and the speed of the engine. Man-B&W port and helix fuel injection pumps have a control sleeve threaded into the lower part of the barrel. When the rack to this control sleeve moves, it raises or lowers the barrel and its ports, thus changing the timing of injection. The plunger also has its own control sleeve and rack to change the effective stroke as noted in the previous discussion of port and helix pumps. Some engines have two racks, one controlling the beginning of injection and one controlling the end. This type of pump has only a lower helix and the top of the plunger is flat.

Timing of injection is very important in producing the maximum amount of power from the fuel.

If the injection timing is too early, the following will happen:

1. Fuel knock. The air will be cooler because of the early injection. This delays the fuel ignition. A larger amount of fuel is injected into the combustion chamber before it ignites, causing the knock.
2. Loss of power. This occurs because the peak pressure is too early and too close to TDC. There is very little moment arm to turn the crankshaft.
3. Noisy engine. The combination of fuel knock and the peak pressure being too high and too close to TDC promotes vibration and noise in the engine.

If the injection timing is too late, the following will happen:

1. Loss of power. The peak pressure is too low (the piston travels downward) and reached too late. The gases are not allowed to fully expand before being exhausted.
2. Hot engine. Because the higher pressures later in the power stroke produce higher temperatures, more heat is transferred to the engine instead of producing power.
3. Hot exhaust. Again, the temperatures later in the power stroke are higher, producing higher exhaust temperatures.
4. Smoky engine. Because the fuel is injected late, some of it does not burn completely, producing smoke.

As you can see, if the injection is too early or too late, a loss of power, and, thus, loss of efficiency occurs.

The engine analyzer or engine indicator can be used to check that the pressure is correct in the cylinder at the proper time. The pressure time trace is needed to do this.

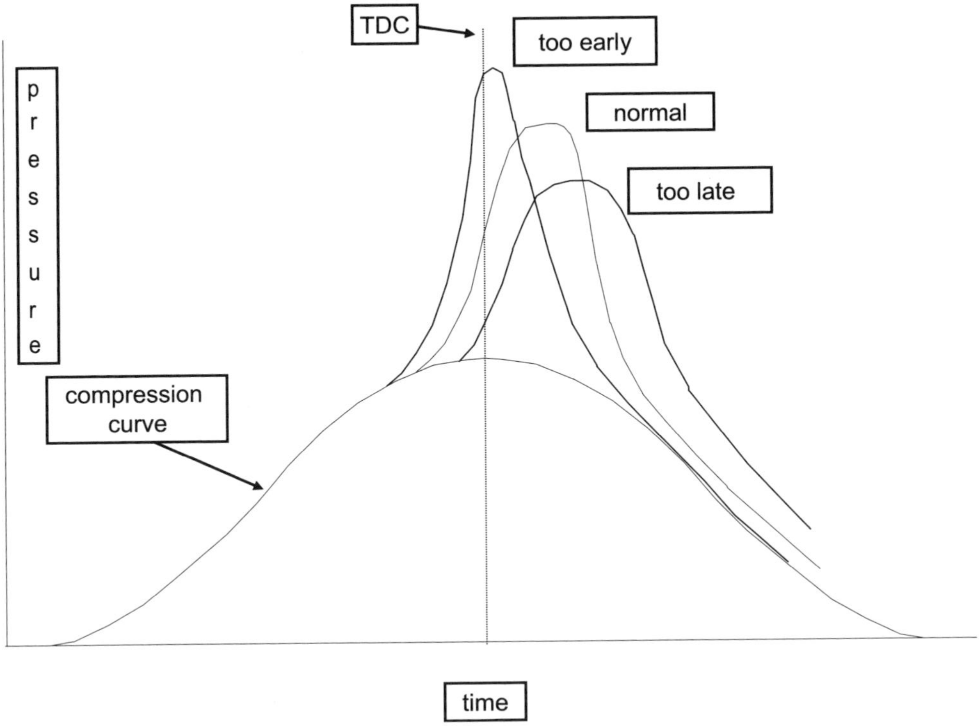

Fig. 11.8. Fuel injection timing

Small Engine Injection Systems

Cummins makes a fuel injection system for their four-stroke engines that is different from other manufacturers. The injector is a high-pressure fuel pump that is cam operated. This pump has no helix. A variable low-pressure pump controls the amount of fuel.

The Cummins fuel injection system works as follows:

1. The plunger in the injector is in the down or closed position most of the time.
2. The amount of fuel fed to the space below the plunger depends on the pressure output of the variable low-pressure pump and the cross-sectional area of the passage to this space.
3. Just before TDC, the plunger rises and a certain amount of fuel is allowed to flow under the plunger, depending on the pressure level from the variable low-pressure pump. The cam forces the plunger down, closing the low-pressure oil inlet and injecting the fuel into the combustion chamber at a high pressure.

Some small, high-speed engines use a distributor type pump to keep costs down. Only one high-pressure pumping mechanism serves all the injectors. A rotating passage delivers the high-pressure fuel to individual passages leading to each injector.

Bosch uses a single plunger that not only moves up and down to pump the fuel, but rotates to distribute the fuel to the individual injectors.

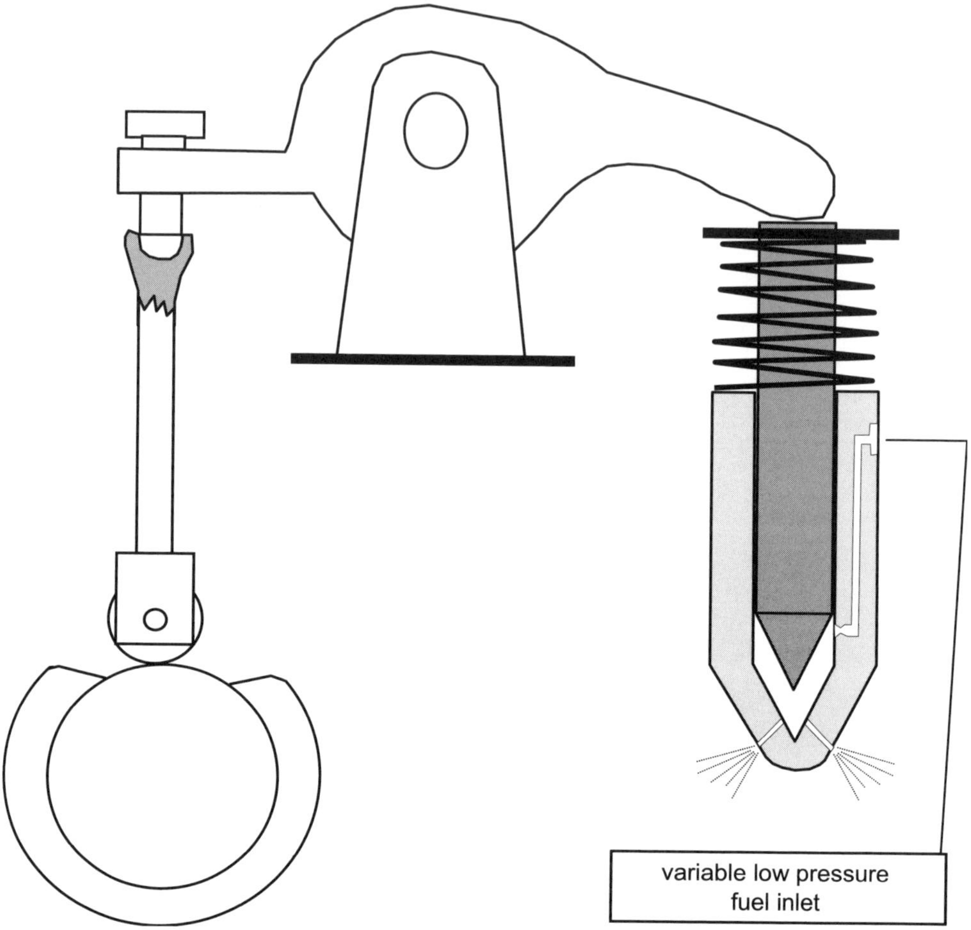

Fig. 11.9. Cummins fuel injection system

Roosamaster uses one set of rotating opposed pistons and an internal cam to pump the fuel. The shaft that rotates the pistons contains an internal passage that distributes the fuel to the individual injector passages.

The unit injector is a rocker arm actuated, high-pressure fuel pump that also contains a differential hydraulic injector. Thus, the injector and a port and helix high-pressure fuel pump are contained in a single unit. General Motors uses this system in most of its high- and medium-speed two-stroke cycle engines.

Differential Hydraulic Injector

The differential hydraulic injector is used with all the high-pressure fuel pumps mentioned. This injector acts much like a safety valve for the high-pressure fuel pump. When the ports close, the fuel pressure increases in the line going to the injector. A spring holds the needle of the injector against its seat. When the fuel pressure reaches a certain point, depending on the compression of the spring, the needle rises, exposing more area to the fuel pressure. The needle pops open wide, allowing the fuel to be sprayed into the combustion chamber. When the ports of the high-pressure fuel pump open, causing the end of injection, the spring forces the needle down. This happens at a lower pressure than the popping pressure because of the larger area exposed to the fuel pressure when the needle lifts. When the needle seats, this area is reduced again, and the original higher pressure is needed to open the injector. This is like a safety valve that has blowdown.

As the needle rises, it increases the volume underneath itself. This lowers the fuel pressure below the needle. The needle is allowed to rise only a fixed amount as it hits the needle stop. A fixed amount of needle lift prevents the needle from reseating.

The nozzle is held by a nozzle holder, which contains the rest of the components to makeup the injector. All sealing is done by ground metal-to-metal surfaces between the nozzle and the nozzle holder. The same close tolerances are used between the needle and the nozzle, just as they are used in the barrel and plunger in the high-pressure fuel pump. These close tolerances contain the high pressures. The needle and nozzle must be replaced as a set, just as the barrel and plunger must be replaced as a set.

The leakoff connection is needed because the needle and nozzle tolerance allows a small amount of fuel to leak into the nozzle holder. This leakage is needed to lubricate the needle and nozzle. Without leakoff connection, eventually, the nozzle holder fills with fuel, causing a hydraulic lock as the needle tries to rise during injection. The leakoff usually goes back into the fuel tank.

The above multiorifice nozzle is used in open combustion chambers. Direct injection is another term for this type of injection system. Because the injector has to mix the fuel with air in the combustion chamber, the popping pressure is very high. At least 3,000 psi is needed for proper atomization, penetration, and mixing. Some newer systems use higher pressures.

The nozzle and nozzle holder shown in fig. 11.11, which makes up the injector, are a basic representation. Some older, heavy-fuel-burning systems incorporate cooling water passages into the injector. Newer models have better cooling passages designed into the heads, eliminating the injection cooling water system.

Another trend in modern, heavy-fuel-burning engines is to circulate fuel through the high-pressure pump and injector at all times except during actual injection. This is accomplished by a shuttle mechanism in the injector that allows fuel to circulate through the injector, and when the pressure coming from the high-pressure injection pump reaches a certain level, the shuttle shuts off circulation. The pressure rises to pop the injector and inject the fuel. This system keeps the fuel at the proper temperature during maneuvering and shutdown of the engine for short in-port stays.

Another type of nozzle used on high-speed engines is the pintle-type nozzle. This nozzle is used with the same type of nozzle holder. It is used with a prechamber that is attached to the main combustion chamber. This prechamber helps the injector mix the fuel with the air. The injector pops at lower pressures (1,200 psi) and produces a coarser spray. Atomization takes place by forcing the fuel through an annulus between the pintle on the needle and the large hole in the nozzle.

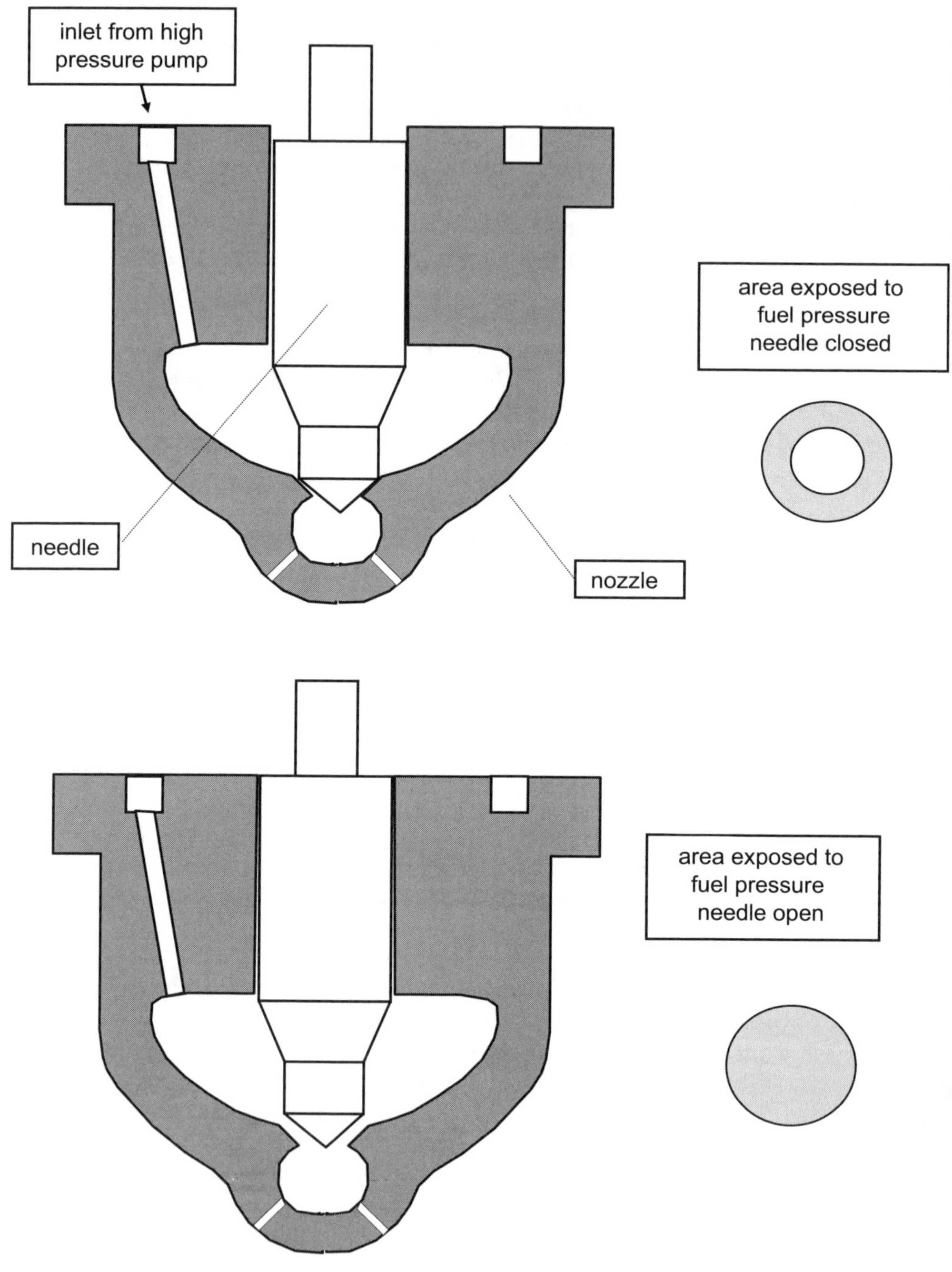

Fig. 11.10. Multiorifice fuel injection nozzle

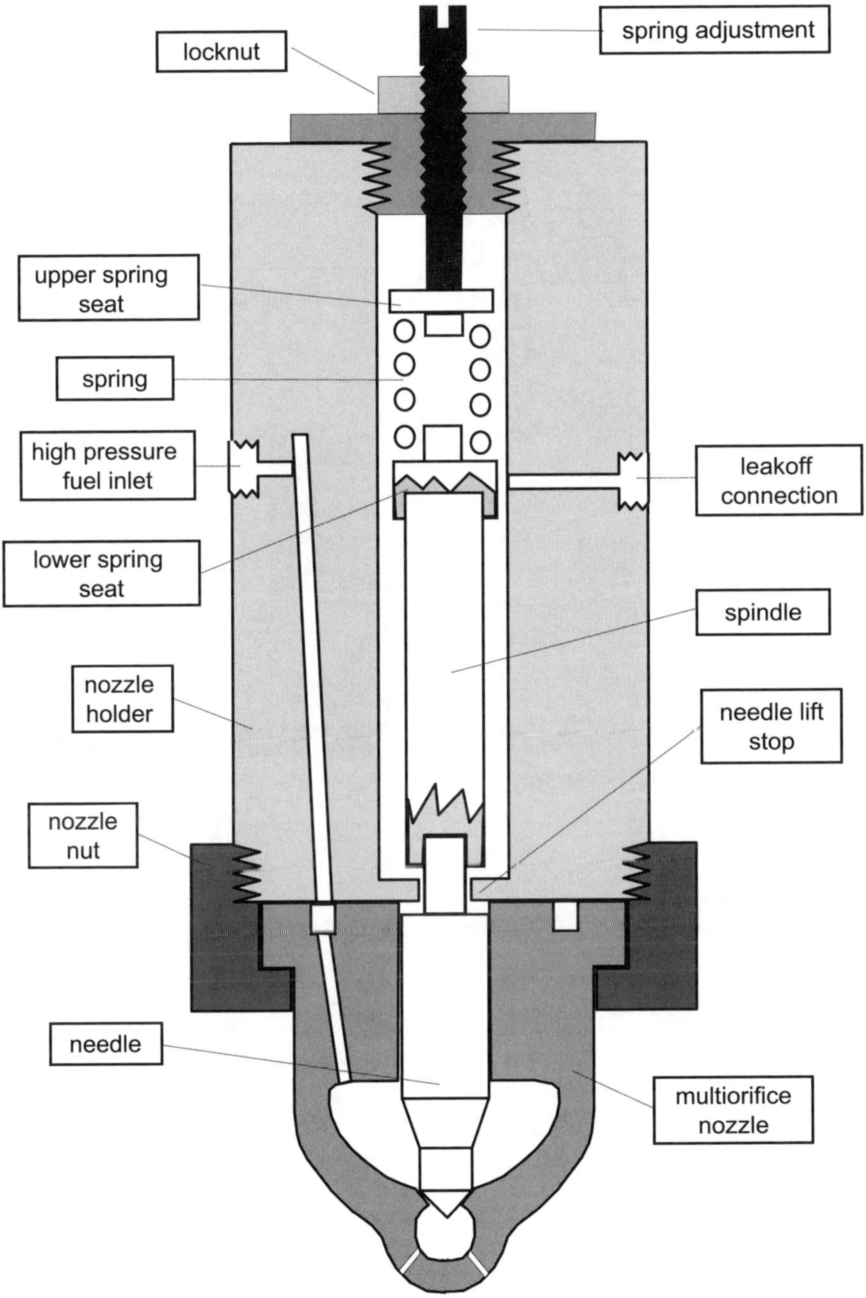

Fig. 11.11. Nozzle holder with multi-orifice nozzle

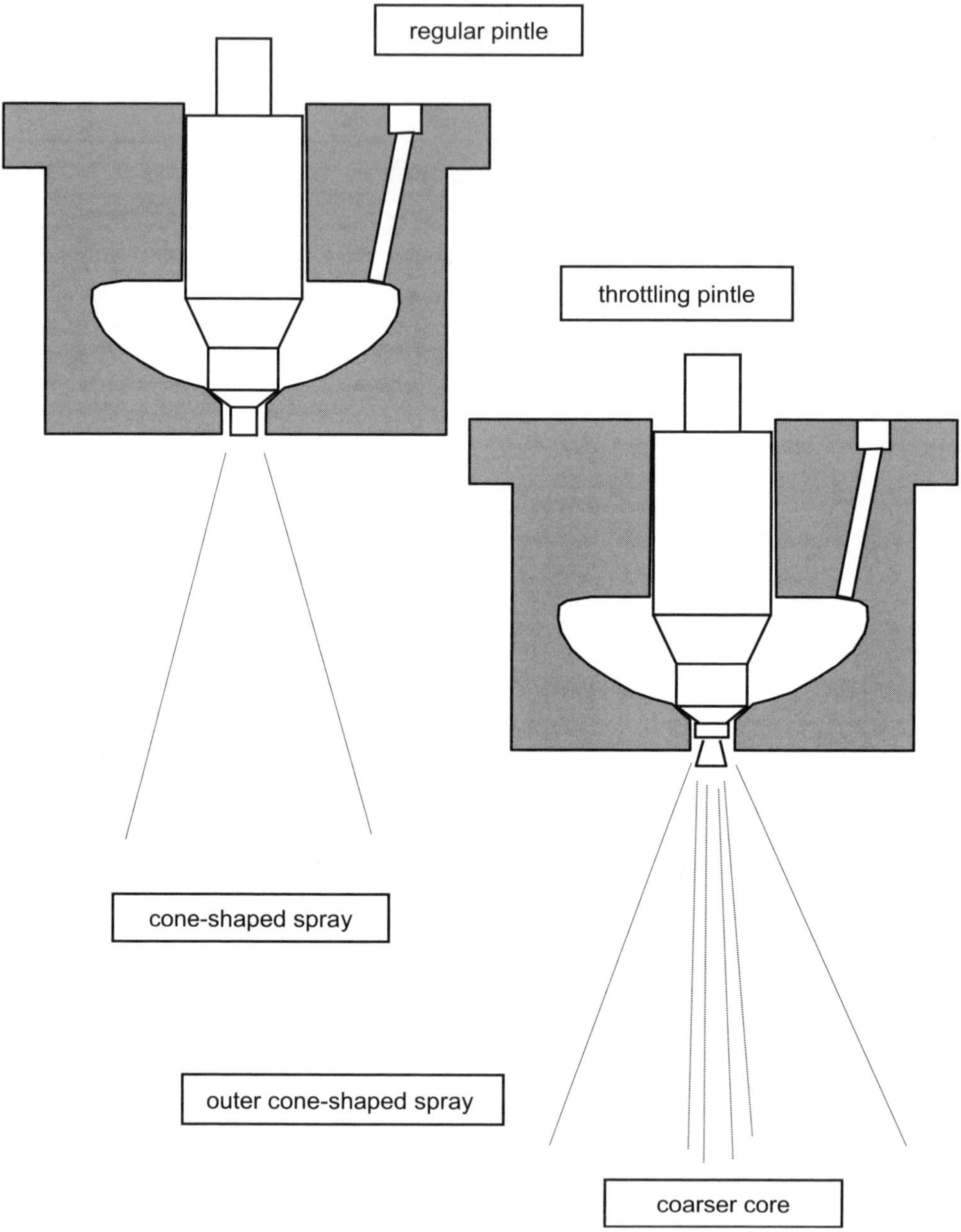

Fig. 11.12. Pintle injection nozzles

In the pintle-type nozzle, the pintle never comes out of the hole in the nozzle because of the needle lift stop. If it did, no annulus would atomize the fuel. The throttling pintle has a two-diameter pintle to develop the cone spray first, and a coarser inner core.

In the design of the fuel injection system, the popping pressure, fuel pressure, needle lift period, and gas pressure in the combustion chamber must be taken into account. The important overriding criteria is that the needle must close while the fuel pressure is still higher than the gas pressure in the combustion chamber. If this is not accomplished, hot gases can get into the injector, carbonizing the fuel around the tip of the needle.

As shown in fig. 11.13, the needle is closed and the fuel pressure is higher than the combustion chamber pressure at the end of injection. You can also see the dip in the fuel pressure as the needle lifts and hits the needle lift stop at the beginning of injection.

The tip of the nozzle also must be kept at the correct temperature.

1. If the temperature is too cold, then moisture may condense on the tip and allow sulfuric acid to form.
2. f the temperature is too hot, the low boiling point lighter fuels vaporize, leaving the heavier fuel components on the outside of the nozzle holes as the fuel is injected. These carbon deposits form carbon trumpets around each one of the orifices on the nozzle. This interferes with the fuel mixing with the air in the combustion chamber. This happens only on medium- and slow-speed, residual-fuel-burning engines.

Older model, heavy-fuel-burning engines had an injector cooling and heating system to control the temperature. More recent designs use cooling passages in the cylinder head near the injector.

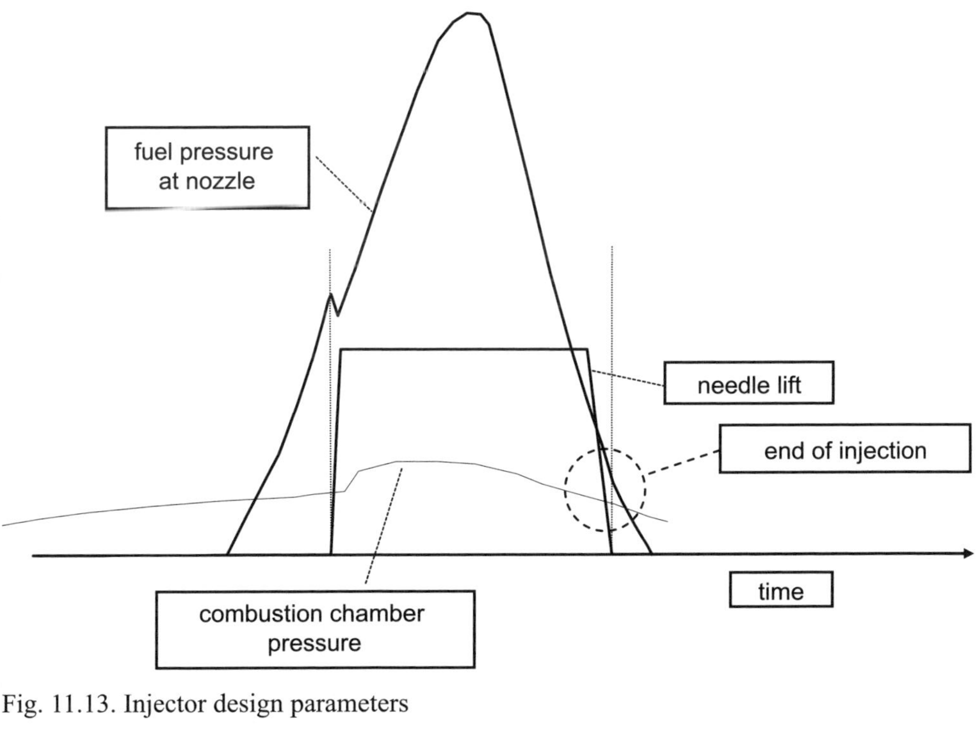

Fig. 11.13. Injector design parameters

In systems with a separate high-pressure fuel injection pump and a separate injector, these items must be connected by high-pressure tubing. This tubing has the following characteristics:

- cold drawn
- soft annealed
- seamless
- SAE 1010 steel
- specially prepared smooth bore

The length of each line from the high-pressure fuel injection pump to the injector is the same length to match the retraction piston.

The tubing is manufactured with integral conical seats. The attachment nuts are placed on the tubing before the seats are made. This means that each set contains the tubing seats and nuts to run the fuel from the high-pressure injection pump to the injector. If any part of the assembly fails, the assembly must be replaced as a set.

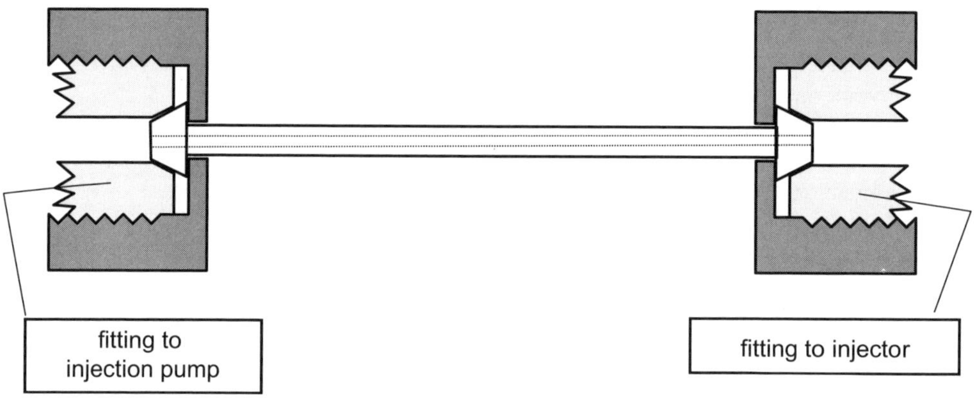

Fig. 11.14. High-pressure fuel injector tubing

Study Questions

1. In a fuel injection system what, ensures that the load in each cylinder is the same?
   A. proper metering
   B. proper timing
   C. proper atomization
   D. proper penetration
2. 2. In a fuel injection system, what ensures the peak pressure is at the proper location?
   A. proper metering
   B. proper timing
   C. proper rate of injection
   D. proper penetration
3. In a fuel injection system, what can determine if the PV diagram looks like an Otto or a Diesel cycle
   A. proper metering
   B. proper timing
   C. proper rate of injection
   D. proper penetration
4. Which fuel injection system most commonly is used on medium-speed diesel engines today?
   A. air blast
   B. common rail
   C. jerk pump
   D. electronic
5. Which fuel system will be most widely used in the future?
   A. air blast
   B. common rail
   C. jerk pump
   D. electronic
6. Which of the fuel injection systems in Question 5 could be part of the same system?
   A. A and C
   B. B and C
   C. B and D
   D. A and D
7. The term effective stroke is associated most closely with which fuel injection system?
   A. air blast
   B. common rail
   C. jerk pump
   D. electronic

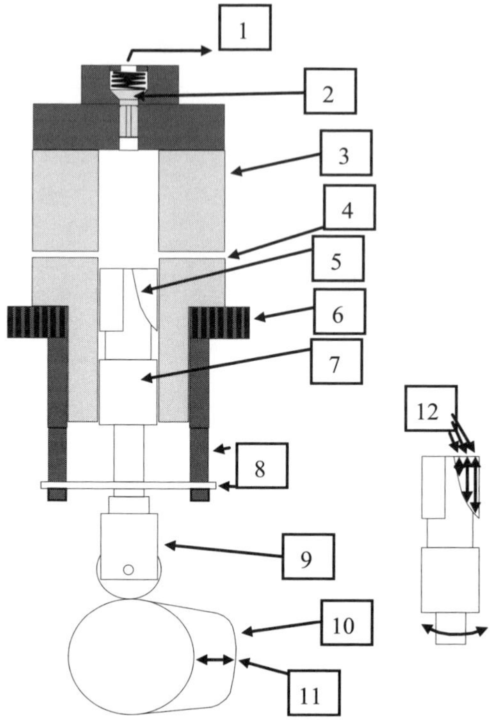

Diagram A

8. In diagram A, the physical stroke is shown by:
   A. 9
   B. 10
   C. 11
   D. 12
9. In diagram A, the fuel rack engages:
   A. 3
   B. 5
   C. 6
   D. 9
10. In diagram A, the effective stroke is shown by:
    A. 9
    B. 10
    C. 11
    D. 12
11. In diagram A, the plunger is shown by:
    A. 3
    B. 4
    C. 6
    D. 7

12. The pump shown in diagram A is a:
    A. common rail pump
    B. port and helix pump
    C. distributor pump
    D. cummins pump
13. The shock wave effect in a high-pressure fuel pump is reduced by:
    A. retraction piston
    B. guides
    C. ports
    D. angled seat
14. Cam surfaces must be designed for which stress?
    A. ultimate
    B. yield point
    C. proportional limit
    D. fatigue
15. If a diesel engine is running with the timing too advanced, the:
    A. exhaust temperature will be higher than normal
    B. engine will smoke
    C. engine will knock
    D. engine will run quietly
16. What opens a high-pressure fuel injector?
    A. pushrod
    B. fuel pressure
    C. camshaft
    D. lube oil pressure
17. A pintle nozzle is used with which type of combustion chamber?
    A. open
    B. prechamber
    C. post chamber
    D. closed
18. Carbon trumpets form when the component they are associated with is:
    A. too hot
    B. too cold
    C. worn
    D. cracked
19. In a fuel injection system, what ensures that the fuel does not hit the combustion chamber walls?
    A. proper metering
    B. proper timing
    C. proper atomization
    D. proper penetration

20. In a fuel injection system, what ensures that the injected fuel has the proper surface area?
    A. proper metering
    B. proper timing
    C. proper atomization
    D. proper penetration
21. Which fuel injection system is most commonly used on medium-speed diesel engines today?
    A. air blast
    B. common rail
    C. jerk pump
    D. electronic
22. In the design of the fuel cam the ports are closed in an area of constant:
    A. displacement
    B. velocity
    C. acceleration
    D. movement
23. Adding shims under the high-pressure fuel pump:
    A. advances the timing
    B. retards the timing
    C. increases the fuel per stroke
    D. decreases the fuel per stroke
24. Which fuel injection system has the port and helix pump and the injector in the same body?
    A. unit injector
    B. Bosch System
    C. Cummings System
    D. jerk pump
25. What prevents an injector needle from reseating?
    A. retraction piston
    B. spring
    C. needle lift stop
    D. spindle
26. A cracked high-pressure fuel injection line must be:
    A. welded
    B. brazed
    C. silver soldered
    D. replaced

Chapter 12
Combustion of Fuel

Combustion in the Diesel Engine

The fuel burned in a diesel engine is a hydrocarbon. C_aH_b is the generic symbol of the molecules that makeup a hydrocarbon. The a and b can be many numbers, depending on the type of hydrocarbon. The fuel is refined from crude oil, using a fractional distillation process. The crude oil is heated and the distillate condensed at different temperatures. The lowest temperatures produce gasoline, a higher temperature kerosene, and so on. The highest temperatures distill heavy lubricating oil. What is left after the distillate is boiled off and condensed is residual oil. This product is of low value and made up of heavy molecules such as asphalt. It can be cracked thermally to produce smaller molecules, such as kerosene or gasoline, to make a more valuable product. The residual also can be cracked catalytically, producing even a larger amount of more valuable product. The catalyst used is aluminum. To have the aluminum present the largest area to the oil, it is used to coat a silicon dioxide substrate (cheap pea stone). As a result of vibration and wear in the cracker, particles of aluminum oxide and silicon dioxide are left in the residual. These are known as catalytic cracking fines or CAT fines. This residual is almost solid and some type of solvent must be mixed in to keep this product a liquid. The particles mentioned above are abrasives, and are used on sandpaper. This fuel must be heated and purified by centrifuges and filters known as CAT fine filters before it can be used as fuel in diesel engines.

The impurities left in the residual would be fine if the residual were discarded or used to pave roads. Since it is still a hydrocarbon it also has value as a fuel.

When the hydrocarbon is burned with the oxygen in the air, the products of combustion are carbon dioxide and water. Nitrogen is in the air, which may be partially oxidized, forming nitrogen oxide. Some oxygen is left over, since the diesel engine runs with a large amount of excess air. Very little carbon monoxide is produced because of the large amounts of excess air. Heavy fuel may also contain sulfur, which burns to form oxides of sulfur. Oxides of nitrogen also are present because of the high temperatures and pressures involved. Oxides of sulfur and nitrogen are considered pollutants. Engine manufacturers have reduced these oxides of nitrogen by modifying the combustion chamber design to comply with international pollution regulations.

A chemical equation for the combustion process is:

$$C_aH_b + O_2 + N_2 + S \Rightarrow CO_2 + N_2 + NOx + O_2 + SOx + H_2O$$

This is not a balanced chemical equation, but it shows how fuel is burned in air and the products of combustion produced. These products of combustion are regulated by international treaty for seagoing vessels. In an effort to reduce oxides of sulfur, regulations under the International Convention for the Prevention of Pollution from Ships (MARPOL) the maximum allowable sulfur content of fuel is 4.5 percent. In Oxides of Sulfur (SOx) emission control areas, such as the Baltic Sea, the limit is 1.5 percent.

The MARPOL limits for NOx content for engines more than130 kW are as follows:

- Engines running at less than 130 rpm – 17g/kWh
- Engines running at 130 rpm, but less than 2,000 rpm – 45 X $n^{-.2}$ g/kWh Where *n* is the rpm
- Engines running at or above 2,000 rpm – 9.8 g/kWh

Local areas are establishing their own limits. Los Angles may require ships in port to shutdown and use shore power.

Modification of the combustion chamber design is not enough to meet some of the new standards on oxides of nitrogen in different parts of the world. Selective Catalytic Reduction (SCR) systems are used on some ships. Ammonia is injected into an exhaust SCR unit that contains a catalyst. The reaction equations below show how the oxides of nitrogen are turned into water and nitrogen gas.

$$4NO+4NH_3+O_2=4N_2+6H_2O$$
$$2NO_2+4NH_3+O_2=3N_2+6H_2O$$

Engine manufacturers also reduce oxides of nitrogen by using water injection. This is done by adding water to the fuel with a homogenizer system just before the high-pressure fuel pump. The water must be in the form of microscopic drops to pass through the holes in the fuel injector and help atomize the fuel and cool the combustion chamber. The explosive boiling of the water droplets inside the fuel drops promotes better and smokeless combustion. The steam also cools the combustion gases, producing less oxides of nitrogen.

The latest development in reducing oxides of nitrogen is injecting high-pressure water into the combustion chamber with a water injector passage in the fuel injector. This is used instead of the homogenizer system. The water and fuel injection are controlled by solenoids, which are part of a computer-controlled injection system.

The EcoSilencer uses a seawater scrubber to reduce SOx by 90 percent and to remove NOx and soot. The seawater is passed through a water treatment system to provide environmentally safe discharge water. Contaminants removed from the seawater are stored for shore disposal.

The stoichiometric air fuel ratio for a typical hydrocarbon is 14.5/1 by weight. This can vary with the type of fuel.

When fuel is mixed with air and brought into the combustion chamber, as in a gasoline engine, the air-fuel ratio must be around this value (14.5/1). A certain range of flammability of the air-fuel ratio must be met or the air-fuel mixture will not ignite. This must be around the stoichiometric air-fuel ratio. If the mixture is rich, it has too much fuel. If the mixture is lean, it has too little fuel. Either way, the fuel will not ignite by the spark in a spark ignition engine or the heat of compression in a compression ignition engine.

The diesel engine operates under a completely different system, being a fuel-injected compression-ignition engine. The air-fuel ratios in a diesel engine can vary from 20 to 1 in a fully loaded engine to 100 to 1 in an idling engine, out of the range of flammability.

Looking at fig. 12.1, you can see that somewhere in the combustion chamber there is a location that is in the range of flammability.

Millions of fuel droplets are injected into the combustion chamber. Near the droplet the mixture is too rich, far from the droplet, the mixture is too lean. Somewhere in between, the mixture is just right to ignite the fuel. The ignition occurs throughout the combustion chamber in an uncountable number of points. Once these flame fronts start, all the fuel molecules are burnt if they are near an oxygen molecule, even if the air-fuel ratio is not in the range of flammability.

Smoke in Diesel Engine Exhaust

Smoke is an indication of whether proper combustion is taking place. The stack of a diesel ship should be clear. This does not negate the requirement of soot blowing. Minute particles of soot accumulate on the finned tubes of the waste heat boiler. This accumulation potentially can start a stack fire. Some ships blow tubes every hour or so with an Infra-

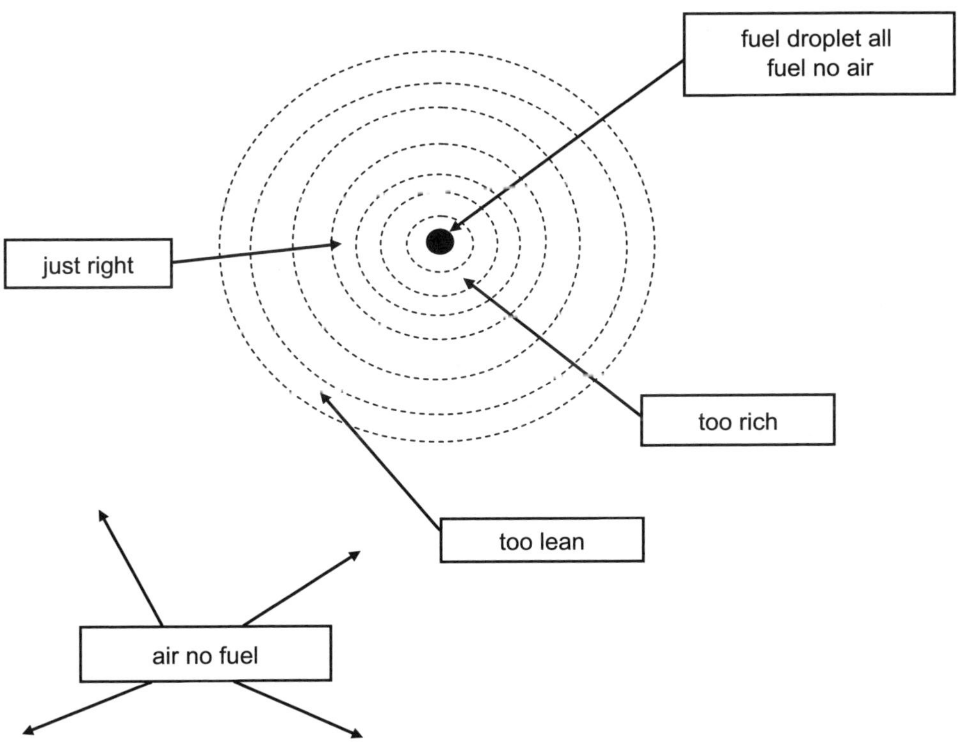

Fig. 12.1. Air-fuel ratios

phone, which is a modified ship's whistle tooted inside the waste heat boiler to shake off soot. Conventional soot blowers must be used in addition to an Infraphone.

Black smoke is an indication of:

- Faulty fuel injection equipment. The fuel is not being atomized and mixed properly with the air.
- Overloaded engine. More fuel than the engine can handle.
- Not enough air. This can be caused by dirty air filters on the turbocharger or engine room ventilation fan inlets, or running too many engine room exhaust fans and not enough supply fans.
- Late timing. The fuel is injected too late to burn completely.
- White smoke is an indication of:
- Faulty injectors. The fuel is hitting part of the combustion chamber or exhaust surfaces and is being explosively boiled, forming a white oil mist.
- Cracked cylinder liner. Cracked head. Cracked tube in waste heat boiler. Water vapor is exiting the stack and condensing.

Blue smoke is an indication of:

- Lube oil being burned. In a trunk piston engine, this indicates of worn piston rings, cylinder liners, valve stems, and guides. In a slow-speed engine, it indicates too much cylinder oil.

Stages of Combustion

In a diesel engine, certain stages of combustion can be seen in a pressure time diagram.

To start injection, the ports or the valves (Sulzer) in the fuel injector pump close. Because the fuel actually is slightly compressible and the high-pressure fuel line is slightly flexible, a delay occurs before fuel is injected into the combustion chamber. This is called injection delay.

1. Injection Begins. The fuel is entering the combustion chamber, but has not ignited. The slight rise in pressure on the diagram is exaggerated greatly. The period between the fuel entering the combustion chamber and the fuel igniting is called ignition delay.
2. Combustion Begins. The fuel begins to burn in a rapid, uncontrolled combustion. This is because the fuel has entered the combustion chamber during ignition delay without igniting and burning. A significant amount of fuel that is rapidly burning is causing a steep rise in pressure.
3. Controlled Combustion. The fuel burns as it enters the combustion chamber, since the fuel in the combustion chamber is burning already.
4. End of Injection. Since no more fuel is entering the combustion chamber, some afterburning occurs and the pressure rapidly drops as the piston goes down from TDC.

Designing the High Pressure Fuel Injection System

This system is designed to admit fuel into the combustion chamber in such a way that the above trace is obtained. The rate of pressure rise should be kept less than 80 psi/degree of crank travel. This is to prevent:

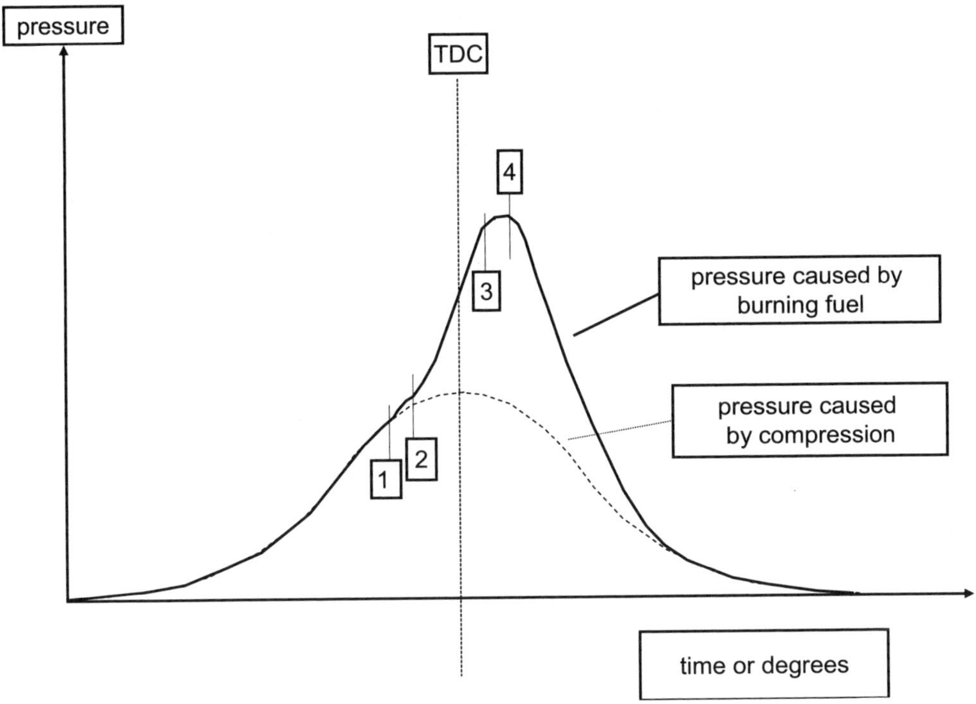

Fig. 12.2. Pressure time diagram

- abnormal engine noise
- roughness in engine operation
- longitudinal vibration and stresses in the connecting rod and crankshaft

Variables in High-Pressure Injection System Design

1. Injection pressure. This is chosen to obtain the correct amount of atomization and penetration.
2. Injection timing. This is chosen to place the peak pressure in the correct position after TDC.
3. Injection duration. This is setup to obtain the correct shape of curve for best engine efficiency.
4. The rate of fuel injection. The rate the fuel is injected into the combustion chamber depends on the size of the holes in the injector nozzle, the plunger diameter of the high-pressure fuel pump, and the shape of the fuel pump cam. These all must be coordinated so that the fuel is injected without blowing the tip off the injector. The duration and the rate of injection go hand in hand. The rate of injection can make the pressure curve look like an Otto cycle or a Diesel cycle. If the fuel is injected as rapidly as possible, the curve looks like an Otto cycle. If the duration is long, then the curve looks like the Diesel cycle.

5. Type of nozzle and combustion chamber. The most efficient combustion chamber is the open combustion chamber, using a multiorifice nozzle injector. However, high-speed engines require some type of prechamber using a pintle-type injector because of the short combustion time. This is the only way to burn fuel with a completely mechanical system in such a short period. Today, using electronically controlled fuel injection systems, an open combustion chamber can be used to burn the fuel in a short period.

One of the biggest considerations in injecting and burning the fuel is the fuel itself. The ignition delay of the fuel is the most important characteristic. Residual fuels have large ignition delays and take time to burn, so they are used in slow- and medium-speed engines. Distillate fuel, such as No. 2 diesel fuel, has a very short ignition delay and is used in high-speed engines and as part of blended fuel in medium-speed engines. A large ignition delay means more uncontrolled combustion and a noisy engine.

Factors Affecting Ignition Delay

1. Centane number. Fuel with a high cetane number decreases the ignition delay. Sixty-fifty is the highest number obtainable with No. 1 diesel fuel.
2. Compression pressure. A high compression pressure or compression ratio will cause higher temperatures and decrease ignition delay.
3. Injection timing: premature injection increases ignition delay.
4. Water jacket temperature. If the jacket water is too cool, this increases ignition delay. This is one reason why a cold engine sounds noisy.
5. Air inlet temperature. If this is too cool, ignition delay is increased. This is another reason why a cold engine is noisy. As the engine warms up, so does the intake manifold and the intake air.
6. Increased rpm. As the speed increases, the fuel injection pumps are operating at their designed speed, causing the fuel to be atomized correctly. Mixing the fuel with the air causes more turbulence in the combustion chamber. This will all decrease ignition delay. This is why an idling diesel engine is noisy. Common rail electronic fuel injection systems overcome this problem by having the same high pressure available at all speeds. However, too much turbulence can thin the boundary layer at the walls of the combustion chamber. This promotes more heat transfer and cool the combustion chamber and increases ignition delay.
7. Air-fuel ratio. The higher the air-fuel ratio, the longer the ignition delay. This is because too much nitrogen in the air is available and tends to cool the combustion chamber, increasing ignition delay. The highest air-fuel ratios are at idle, thus adding another factor of noisy and bad combustion while idling.

Combustion Chamber Design and Air Movement

As the fuel is injected into the combustion chamber, fresh air must be near the fuel to promote combustion. As the fuel burns, the products of combustion must be taken away from the entering fuel.

This is done in two ways:

1. The velocity of the atomized fuel propels it through regions of fresh air in the combustion chamber. The higher the injection and popping pressure, the more the fuel is propelled through the air for better mixing.

2. The combustion chamber can help mix the fuel and air. This can be done by angled, inlet ports on two-stroke cycle engines and angled, inlet passages on four-stroke cycle engines. These are general cases and different styles of combustion chambers promote other ways of mixing fuel and air.

Mixing fuel and air is done by creating turbulence in the combustion chamber. There is a danger of creating too much turbulence, which can decrease the thickness of the boundary layer of the air at the walls of the combustion chamber during intake or scavenging. This cools the air charge and increases ignition delay.

Large main propulsion engines use the open type of combustion chamber. Slow-speed engines of today are uniflow scavenged so they have multiple injectors located around one centrally located exhaust valve. Medium- and high-speed two-stroke cycle engines have one centrally located injector with multiple exhaust valves. Four-stroke cycle engines have a centrally located injector between intake and exhaust valves. Multiorifice nozzles with high-fuel injection pressures are used to aid penetration and mixing of fuel and air. The chamber uses squish areas to create air flow in the combustion chamber independently of the injector. This does not create a large amount of turbulence in the combustion chamber. The injector and the high injection pressures mainly cause the mixing in an open combustion chamber.

Medium- and slow-speed engines use open-type combustion chambers. Ship's service diesel generators also used this type of chamber. This chamber is the most efficient configuration in use. Engines using this configuration have the highest thermal efficiency. Because of the limited amount of turbulence in this chamber, it has been used on slower engines. The less mixing of the fuel and air, the longer combustion takes. The trend is to try and use this configuration for high-speed engines also. To do this, extremely high injection pressures are needed to mix quickly fuel and air to promote rapid combustion. Not until the use of computer-controlled systems has this become possible. All high-speed engines, whether they use prechambers or a computer-control system, must burn No. 2 fuel oil because of the short burn time.

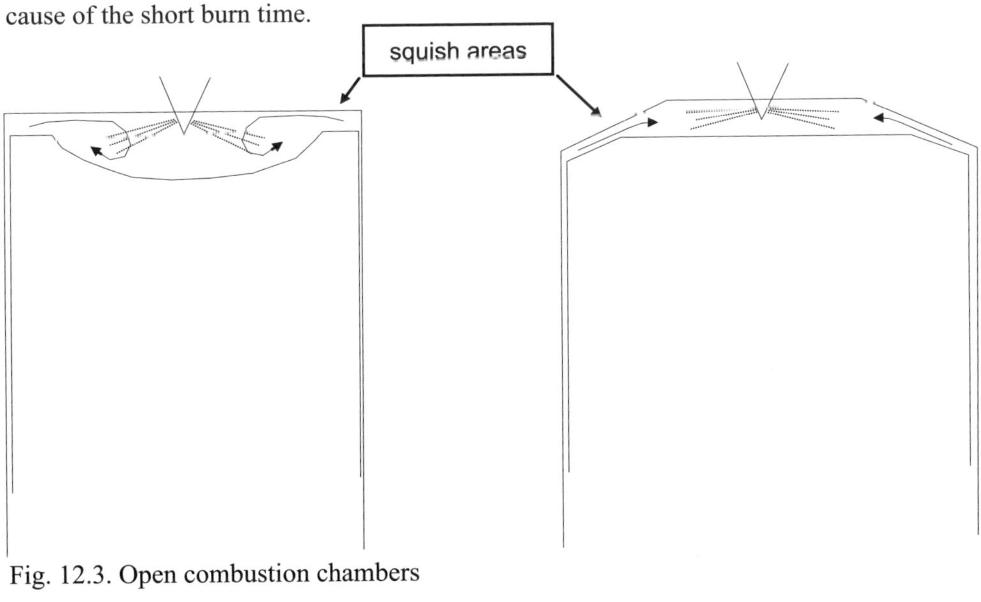

Fig. 12.3. Open combustion chambers

The use of a prechamber with a pintle-type injector nozzle is another solution to mixing fuel and air for rapid combustion in a high-speed engine. With this configuration, the throttling of the gases between the prechamber and the main chamber does most of the mixing. The pintle nozzle produces a coarser spray operating at lower injection pressures. The lower pressures are adequate because the prechamber does most of the mixing. The lower pressures in the fuel injection system allow a less expensive system to be produced.

Other advantages are:

1. Lower peak pressures in the main combustion chamber.
2. Smoother combustion in the main combustion chamber.
3. Less sensitivity to variations in the quality of the distillate fuel used.
4. Prechamber can be made of high-alloy, heat-resistant steel to retain heat to help ignite the fuel. The fuel spray can even impinge on a hot spot within the prechamber.

Disadvantages to prechambers are:

1. It is hard to remove exhaust gases from the prechamber, thus decreasing volumetric efficiency.
2. A glow plug is needed on some prechambers for starting. A glow plug is an electric resistance heating element that glows red to help ignite the injected fuel only during starting.
3. The engine has a lower thermal efficiency due to:
    - Increased overall area of combustion chamber promoting heat loss to the cooling water.
    - Increase in turbulence, thus promoting more heat transfer to the cooling water.
    - Throttling loss from prechamber to main combustion chamber, which is dissipated as heat to the cooling water.

The prechamber is used because it is the only available, completely mechanical design with a proven track record for high-speed engines. It is now being replaced by computer-controlled electronic systems.

The prechamber can be designed as a sphere or a cylinder. The sphere is the most efficient shape to contain pressure. The cylinder is the second-most efficient shape to contain pressure. Because of this, the prechamber can be designed to be fairly light weight and

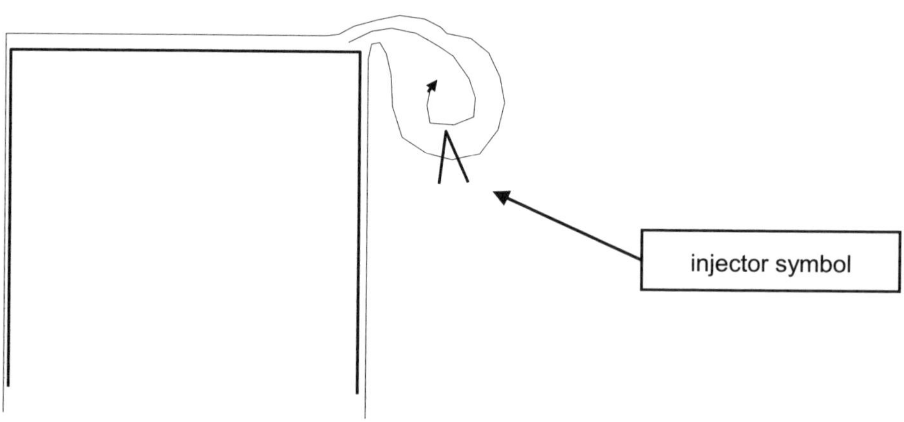

Fig. 12.4. Turbulence chamber

contain very high peak pressures. The pressures can be throttled to the main combustion chamber to reduce the peak pressures on the cylinder head, piston, and liner. Thus, the engine can be made much lighter. We are trading some inefficiency for less weight.

The turbulence chamber contains 80 to 90 percent of the clearance volume. The regular pintle injector is used with this chamber. Most of the combustion takes place within the chamber and is throttled out to the main chamber above the piston. This is the most inefficient design but the engine can be made the lightest.

The precombustion chamber contains 35 to 40 percent of the clearance volume. The regular pintle injector is used with this chamber. Combustion starts in the chamber, but its size does not allow enough oxygen to complete the combustion. To complete combustion,

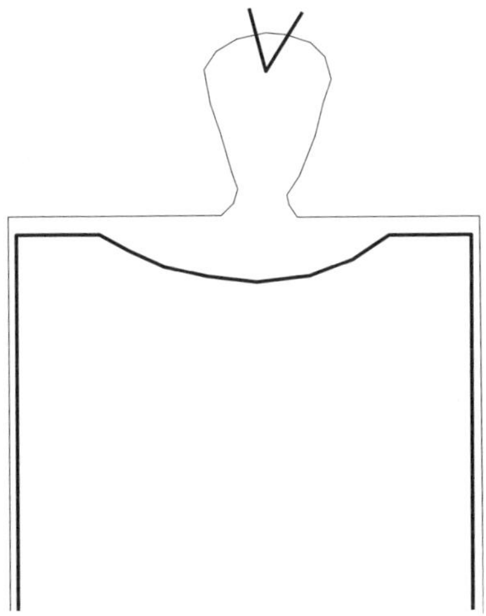

Fig. 12.5. Precombustion chamber

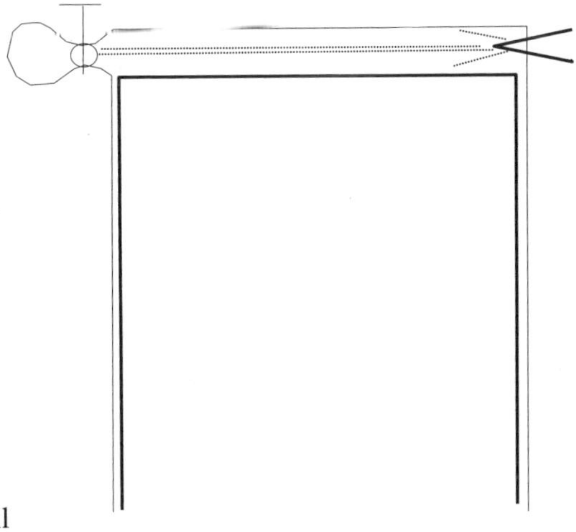

Fig. 12.6. Energy cell

a flaming mass of burned and unburned fuel is blown into the main chamber, mixing with more oxygen. Since the precombustion chamber is smaller, this configuration is more efficient, but requires a more heavily built engine.

Because of their configuration, the above prechambers need glow plugs for starting.

The energy cell contains 10 to 20 percent of the clearance volume. The throttling pintle is used with this chamber. Combustion starts in the main chamber as a result of the initial fine, conical spray. Some fuel is shot into the energy cell by the coarser inner core spray. As it combusts, it is blown back into the main chamber, creating turbulence and mixing the burning fuel. Some designs have a valve shut off the cell to increase the compression ratio for starting. The cell is open during running to promote smooth combustion. Since this chamber is smaller still, the efficiency increases, but the engine is heavier, since more of the peak pressure occurs in the main chamber above the piston.

Fig. 12.7 shows the Lanova Energy Cell. This cell is unique because it contains a double chamber.

The air cell is just a chamber located on the side of the main combustion chamber or in the piston. No combustion takes place in the cell. It is designed to increase turbulence as air rushes out of the cell near the end of the power stroke. This cell is the smallest, thus, this configuration will have the highest thermal efficiency, but will need a heavily built engine to contain the peak pressures in the combustion chamber. A multiorifice injector has to be used with the air cell.

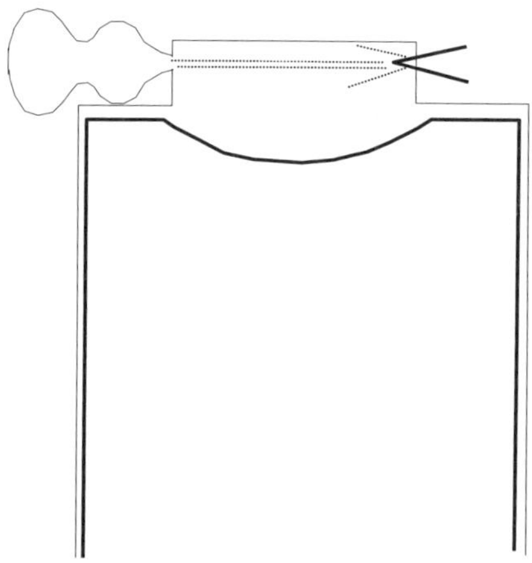

Fig. 12.7. Lanova energy cell

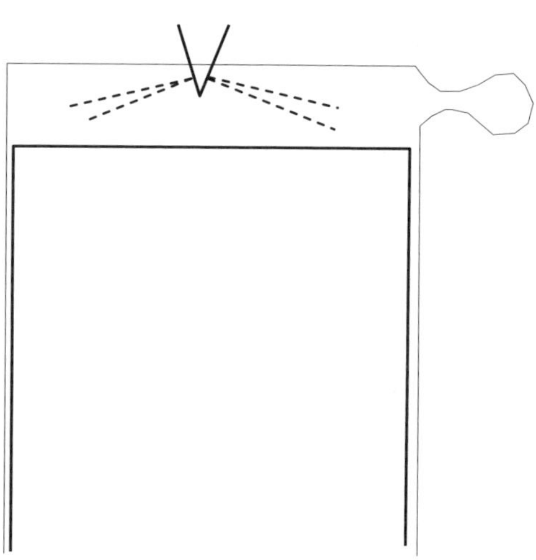

Fig. 12.8. Air cell

Study Questions

1. Cat fines are found in:
   A. No. 2 fuel oil
   B. lubricating oil
   C. residual oil
   D. pneumercator oil
2. Which of the following products of combustion is limited because of international standards?
   A. N_2
   B. O_2
   C. NO_x
   D. H_2O
3. What promotes a short ignition delay?
   A. heavy fuel oil
   B. low intake air temperature
   C. high air fuel ratio
   D. low air fuel ratio
4. The thickness of the boundary layer in a combustion chamber can be decreased by:
   A. too little turbulence
   B. too much turbulence
   C. too high a cooling water temperature
   D. too low a cooling water temperature
5. Which is the least efficient type of combustion chamber?
   A. open combustion chamber
   B. closed combustion chamber
   C. precombustion chamber
   D. turbulence chamber
6. Squish areas are most important in what type of combustion chamber?
   A. open combustion chamber
   B. energy cell
   C. precombustion chamber
   D. turbulence chamber
7. In the fractional distillation process, which boils off at the lowest temperature?
   A. residual oil
   B. lubricating oil
   C. kerosene
   D. gasoline
8. The period between the time the fuel is injected into the combustion chamber and the time it begins to burn is called:
   A. injection delay
   B. ignition delay
   C. impulse delay
   D. reaction delay

9. What promotes a short ignition delay?
   A. heavy fuel oil
   B. low intake air temperature
   C. high air-fuel ratio
   D. low air-fuel ratio
10. Slow-speed engines use which combustion chamber?
   A. open combustion chamber
   B. closed combustion chamber
   C. precombustion chamber
   D. turbulence chamber
11. Glow plugs are used in:
   A. open combustion chambers
   B. starting
   C. emergency lighting
   D. fuel oil service heating
12. Which combustion chamber is associated with the heaviest built engine?
   A. open combustion chamber
   B. energy cell
   C. precombustion chamber
   D. turbulence chamber
   D. turbulence chamber

Chapter 13
Cylinder Heads and Valves

Cylinder Heads

Cylinder heads need many internal passages for the following components:

- intake and exhaust valves for four-stroke engines, only exhaust valves for two-stroke cycle uniflow scavenged engines
- cooling passages
- safety valves, slow-speed engine
- indicator cock, all but high speed
- air start valve, all but high speed
- fuel injector

Casting is the cheapest process that allows us to make such a complicated shape with so many internal passages. Small high-speed engines can use an alloy cast iron. Larger, high-output engines have to increase the thickness of the head to contain the pressure acting on a larger area, which can result in thermal stresses. The head then may have to be made of cast steel to reduce the thickness.

Large slow-speed engines are under great stress due to the large bore area involved. A forged steel head with all passages machined in is required. This includes the bore cooling. This is an expensive process, but it is the only way because of the large forces.

Cylinder Liners

The three basic types of cylinder liners are:

1. Dry liners. These liners are replaceable cylinders pressed into a hole bored into the cylinder block. They do not have direct contact with the cooling water.
2. Integral liners. These liners have a water jacket built into the liner. Only a small inlet and outlet passage needs to be lined up in the cylinder block to allow cooling water to cool the cylinder wall.
3. Wet liner. These liners are in direct contact with the cooling water. The top of the liner has a metal-to-metal seal with the cylinder block, while the lower end uses O-ring seals. Wet liner is used in most modern engines.

The wet liner is used in main propulsion and diesel generator engines. This liner has no thermal barrier as the dry liner does. The wet liner is in direct contact with the cooling water.

Fig. 13.1 depicts a representation of a completely ported, Sulzer slow-speed engine cylinder head and liner. Note the following:

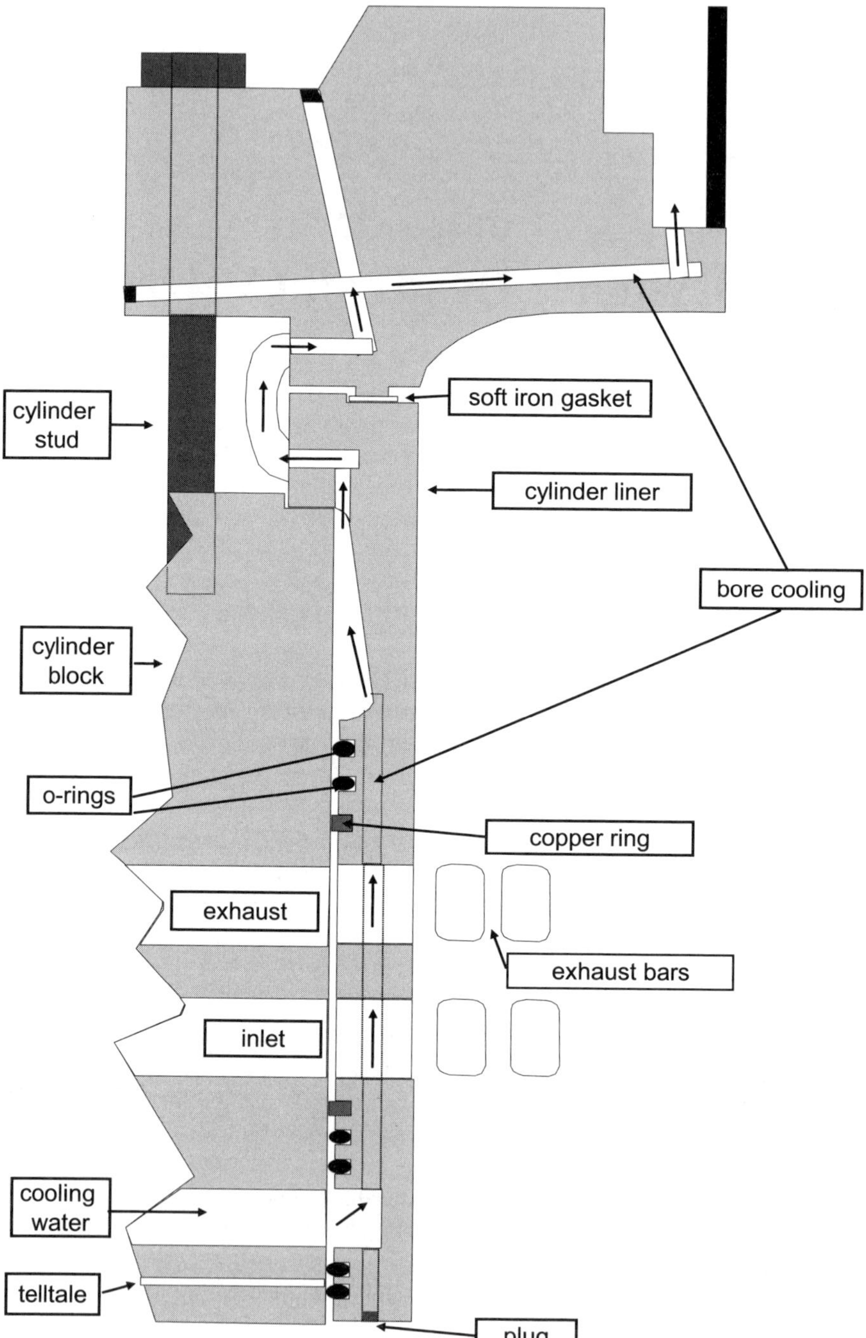

Fig. 13.1. Sulzer slow-speed cylinder head and liner

- The head has a simple shape since it has to be forged. All cooling water passages are machined.
- If the head is over tightened, a bending moment is introduced in the head studs and the top of the liner, probably causing overstress and high-pressure gas leakage. The head and liner are made very thick where they meet to take care of this. The head nuts usually have special spherical washers to prevent any bending of the head to be transmitted to the stud.
- The exhaust bars are drilled for the bore cooling. They also should be checked with a straightedge for wear during cylinder inspection. Most modern engines are uniflow scavenged, and do not have exhaust bars in the liner. Instead, they have an exhaust valve in the head and inlet ports in the bottom of the liner.

Valves

Spring-loaded poppet valves are used to control the flow of the inlet air and exhaust in four-stroke cycle engines. Two-stroke cycle engines may be completely ported in older designs, thus eliminating both intake and exhaust valves. Modern two-stroke cycle engines are uniflow scavenged and have a timed exhaust valve in the head.

Fig. 13.2 illustrates usual valve nomenclature.

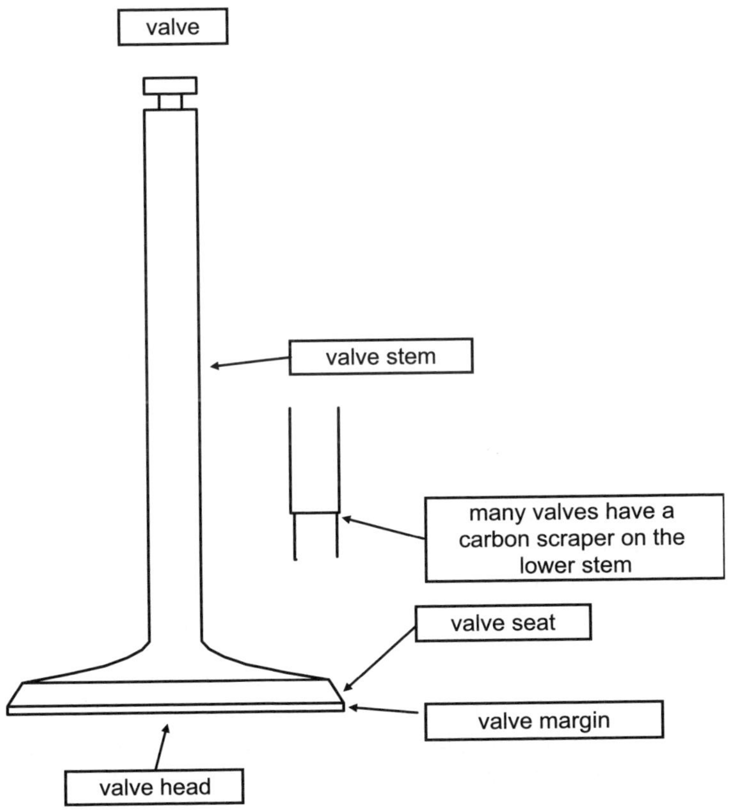

Fig. 13.2. Valve nomenclature

Intake valves do not have a heat dissipation problem, since only cool inlet air flows over the valve head. They can be made of a low-alloy, ferritic steel.

Exhaust valves have a heat dissipation problem because of the hot exhaust gases flowing over the head of the valve. The exhaust valve must be made of a high-alloy, austenitic steel. High-output turbocharged engines usually have the valve seat faced with stellite. This is a cobalt-based steel alloy that is highly heat and abrasion resistant. To save money, some exhaust valves use a lower-alloy steel stem welded to a high-alloy valve head.

Heat itself is not the biggest problem. Elevated exhaust valve temperatures can promote exhaust valve deposits. If a part of the deposits crack off, a gap in the sealing surface is left. This results in a burned exhaust valve. The term used for this phenomena is wire drawing.

The valve is only part of the gas control apparatus. The rest is contained in the head, as shown in fig. 13.3. The valve is held in the closed position by the valve spring. The spring is precompressed to tightly close the valve.

The valve rides up and down in the valve guide. This valve guide usually is replaceable, since it wears out eventually.

Nonturbocharged engines have a valve stem seal to prevent excess oil from entering the valve guide. In turbocharged engines, the seal is not needed because a positive pressure is maintained below the guide.

The stem clearance between the valve stem and the valve guide must be within specifications. Too much clearance allows gas leaks and can pull in too much lube oil. Too little clearance results in a lack of lubrication and a stuck valve.

Fig. 13.3 shows a valve seat insert in the head that contains half a water passage. This means the cooling water is in direct contact with the valve seat insert. The insert is made of stellite. The inserts in older engines did not have this feature. A solid insert without the water ring was pressed into the head. This left a barrier to heat flow between the insert and the metal of the head. This insert is used only on exhaust valves because of the hot exhaust gases that flow around them.

Valve Maintenance

When the valves are removed, they must be checked for the following:

- Valve stem clearance.
- Bent stems. This is done by using a dial indicator while the valve either is chucked in a lathe or resting on V blocks.
- Worn seats.
- The use of a profile tool from the manufacturer to check the seat and margin. The valve is replaced if it does not meet specifications. In the case of worn seats, the valve may be reconditioned. This can be done in a lathe, using a tool post grinder instead of a tool bit. The seat can be ground only to the point that the margin still meets its minimum dimension. This ensures enough heat transfer at the edge of the valve head to prevent distortion and, in the worst case, tuliping.

The seat in the head also can be reconditioned using a special grinding machine that is supported by an arbor in the valve guide. Manufacturers recommend that the seats be ground instead of lapped in. Too much lapping in can mask the valve and restrict air flow.

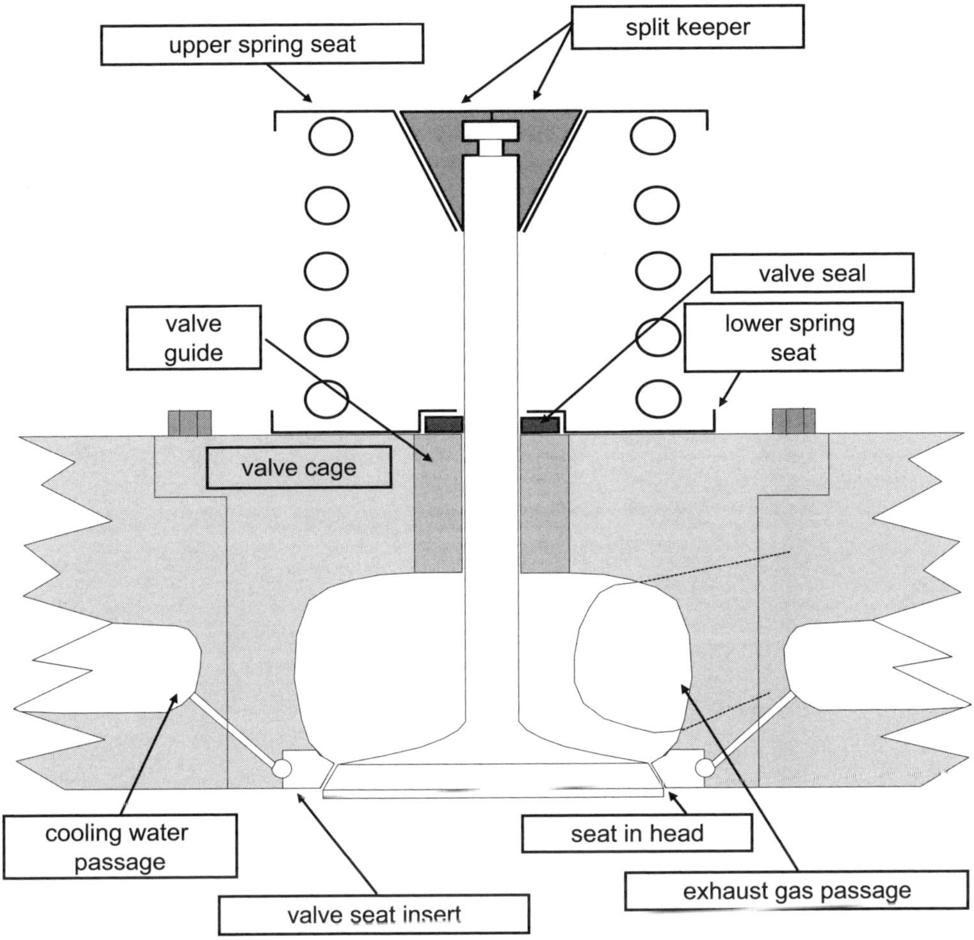

Fig. 13.3. Valve cage

With the material outlined by the dotted line in fig. 13.5 being lapped away, the valve sinks into the head. As the valve starts to open, the lapped away recess essentially stops air flow until the valve opens beyond this area.

To facilitate valve maintenance, most main propulsion engines use valve cages. The whole valve assembly can be removed without taking off the engine head. A new or re-conditioned valve cage can then be fitted and the worn valve and seats reconditioned or replaced without having to hold up the ship for engine repairs.

Valve Springs

Valve springs are another item in the valve cage that must be inspected. The spring should be checked for cracks. The fatigue life of the valve spring also is important, since it goes through so many cycles as the engine is running.

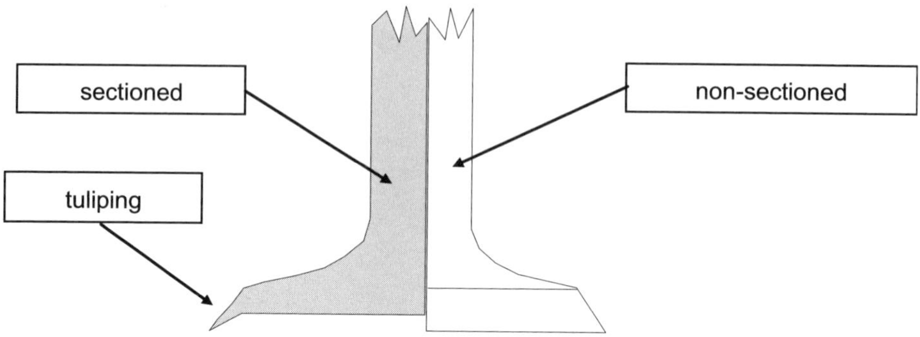

Fig. 13.4. Valve tuliping

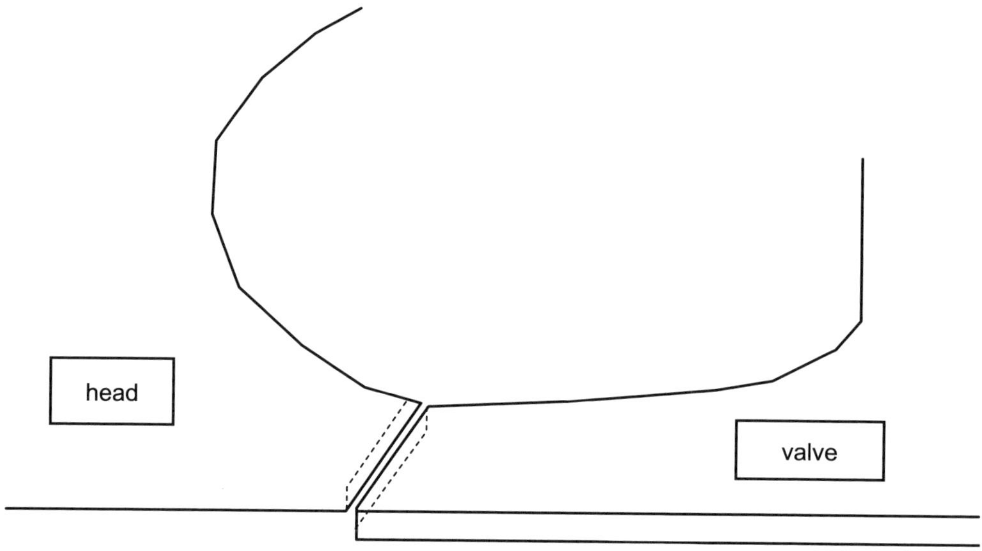

Fig. 13.5. Valve lapping

The spring constant should be checked. The spring constant is checked in a machine that deflects the spring while measuring the force and the length of the spring.

Valve springs must be designed to have a high enough (k) spring constant, so the cam follower will not leave the surface of the cam at any time, or cam follower bounce occurs. This results in an impact load on the cam and follower surfaces.

Cam follower bounce can occur as the follower decelerates near the nose or on the back side of the cam, as the cam is pulling away from the follower.

Harmonics of the natural frequency of the valve spring must be checked. If a wave is traveling along the spring while it is being compressed, part of the spring will be compressed more than it was designed for. This results in a fatigue failure. Some engines use

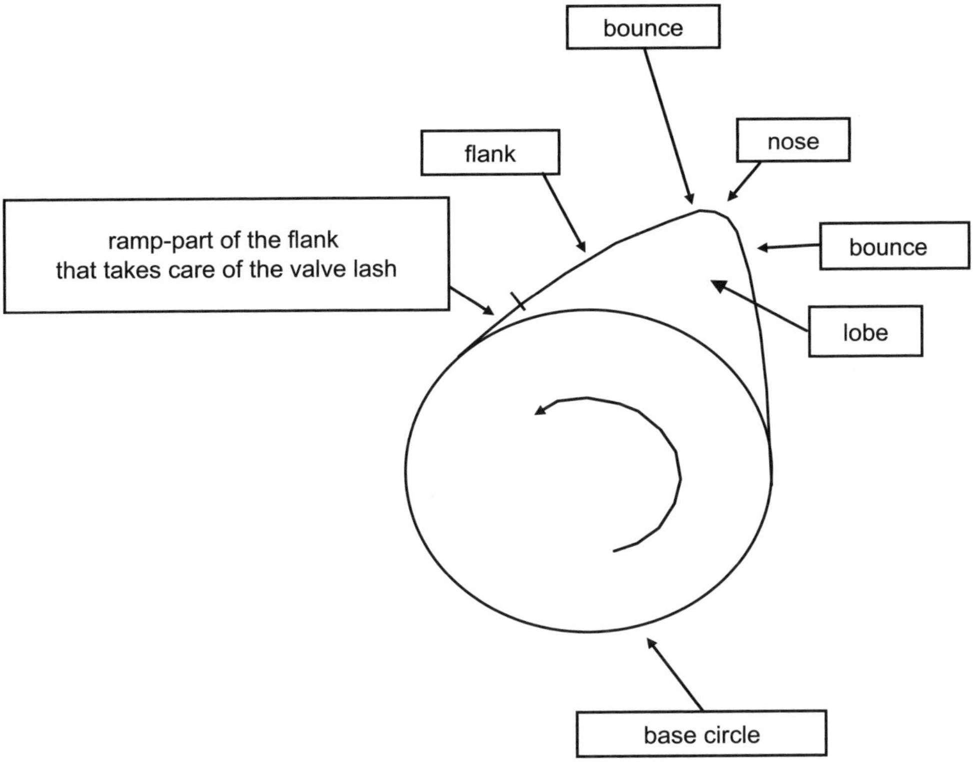

Fig. 13.6. Cam follower bounce

two valve springs per valve to change the natural frequency, while obtaining the proper spring constant.

All valves are cooled through their seats when the valve is closed. The valve seats also are cooled through conduction up the stem to the valve guide. In turbocharged engines, both valves can be cooled by air blowing by them during the overlap period.

Besides the heat of the exhaust gases, exhaust valve seat deposits cause problems, as previously mentioned. The deposits only stick to the valve if the seats are above a certain temperature when burning residual oil. If the seats are below this temperature, the deposits bounce off and continue up the stack as fly ash. The deposits that stick to the valve partially crack off, leaving a gap that hot combustion gases can blow through and erode the valve seat. This is known as wire drawing. By the time hot combustion gas leakage can be detected by the exhaust gas pyrometer, the valve seats have been damaged beyond repair. Methods of providing extra cooling for exhaust valves are listed below:

1. The stellite valve seat insert in the head can be in direct contact with cooling water, as seen in the diagram with the valve cage.
2. Valve rotators can be used to rotate the valve and knock off deposits. Since gas flow is not exactly symmetrical, rotating the valve promotes even heating and less distortion. This prevents hot gases from wire drawing the seats. It also knocks off deposits.

The top valve rotator (fig. 13.7) uses ball bearings and an inclined plane to allow the part attached to the stem to rotate as the valve opens. When the valve is closed, the lash allows the springs to push the ball back up the plane. These balls and inclined planes are incorporated into the upper spring seat.

The rotator in fig. 13.7 uses a spring that falls to one side as the valve is opened. When the valve is closed, the lash allows the spring to pop back up.

The simplest type of valve rotator has angled blades attached to the stem. Enough force is generated to slightly rotate the valve as it opens and closes. This blade-type rotator was mainly used on slow-speed engines, but is coming into use on some medium-speed engines.

3. Sodium-filled valves can use convection currents to move the heat into the valve guide area. Sodium is a liquid at the operating temperatures of the engine.
4. The valves can be water cooled with flexible hoses.

The above are listed in order of reliability. The water cooled valve seats have no moving parts, while the flexible hoses may fatigue and leak. The moving parts of some valve rotators may fail. The sodium-filled valves may break and the sodium react violently with any water in the combustion chamber.

Valve Seat Angles

The magnitude of valve seat angles are important for sealing, gas flow, and pounding.

Larger angles promote better sealing, because the valve jams into the seat. However, this promotes more pounding of the seats because of the jamming action.

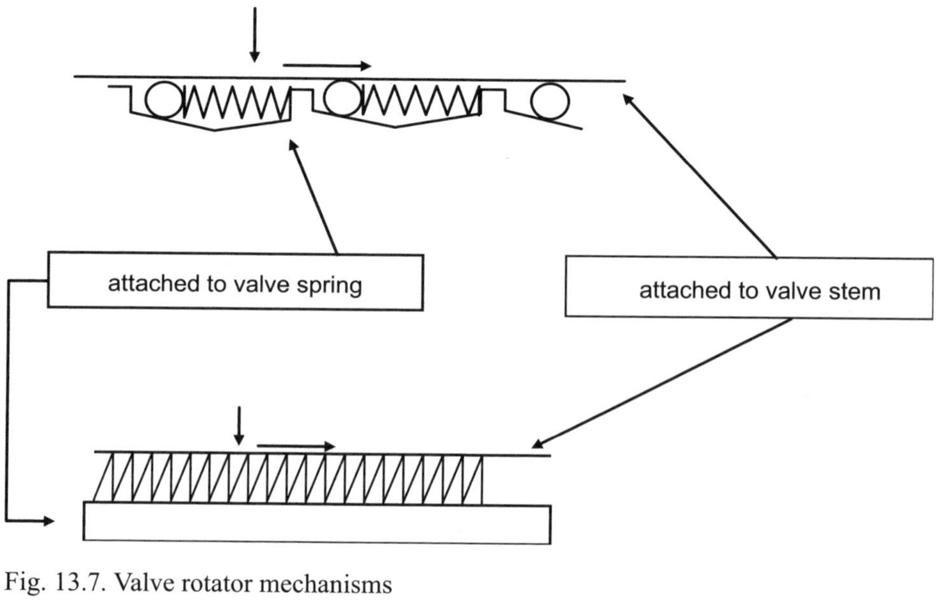

Fig. 13.7. Valve rotator mechanisms

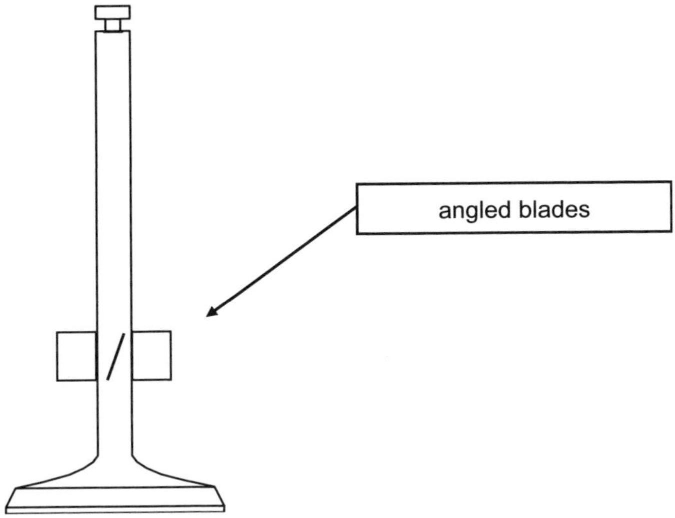

Fig. 13.8. Exhaust gas-operated valve rotator

Smaller angles promote better gas flow, because the area perpendicular to the gas flow is greater with the same valve opening. A valve angle of 30 degrees has a 20 percent greater area perpendicular to gas flow than a valve angle of 45 degrees.

Cosine θ = P/L from this relationship, the dimension the air flow experiences (P) can be calculated from the valve lift (L) and the valve angle θ .

Fig. 13.10 depicts the rest of the valve gear. Most shipboard diesel engines use some type of pushrod type of valve actuation.

Valve lash is the built-in play in the valve actuating system and usually is checked when the engine is cold. It must be within specified dimensions to allow for the thermal growth of the long valve-stem and pushrod as the engine heats up to the operating temperature. The valve when it is supposed to be closed may stay slightly open with too little lash. Too little lash causes the valve to open early and close late. The valve may not open enough with too much lash.

Other valve actuating systems are available for overhead camshaft engines, but they are used on small high-speed engines.

All valve actuating systems require lubricating oil passages for the camshaft, rocker arm, cam follower, and valve guide. This oil usually is supplied by the main engine oil lube service system. Some main propulsion engines have a separate system for the camshaft, rocker arms, and valve guides.

Some small high-speed engines use hydraulic lash adjusters, also called hydraulic valve lifters. They usually are located in the cam follower, but may be located in the rocker arm area. They are essentially a hydraulic piston with a designed amount of leakage between the piston and cylinder. Oil is supplied by a check valve from the oil supply to the camshaft or rocker arm. When the valve is shut, the oil pressure pushes the piston out and takes away any lash. When the valve is opening, some trapped oil leaks out of the hydraulic

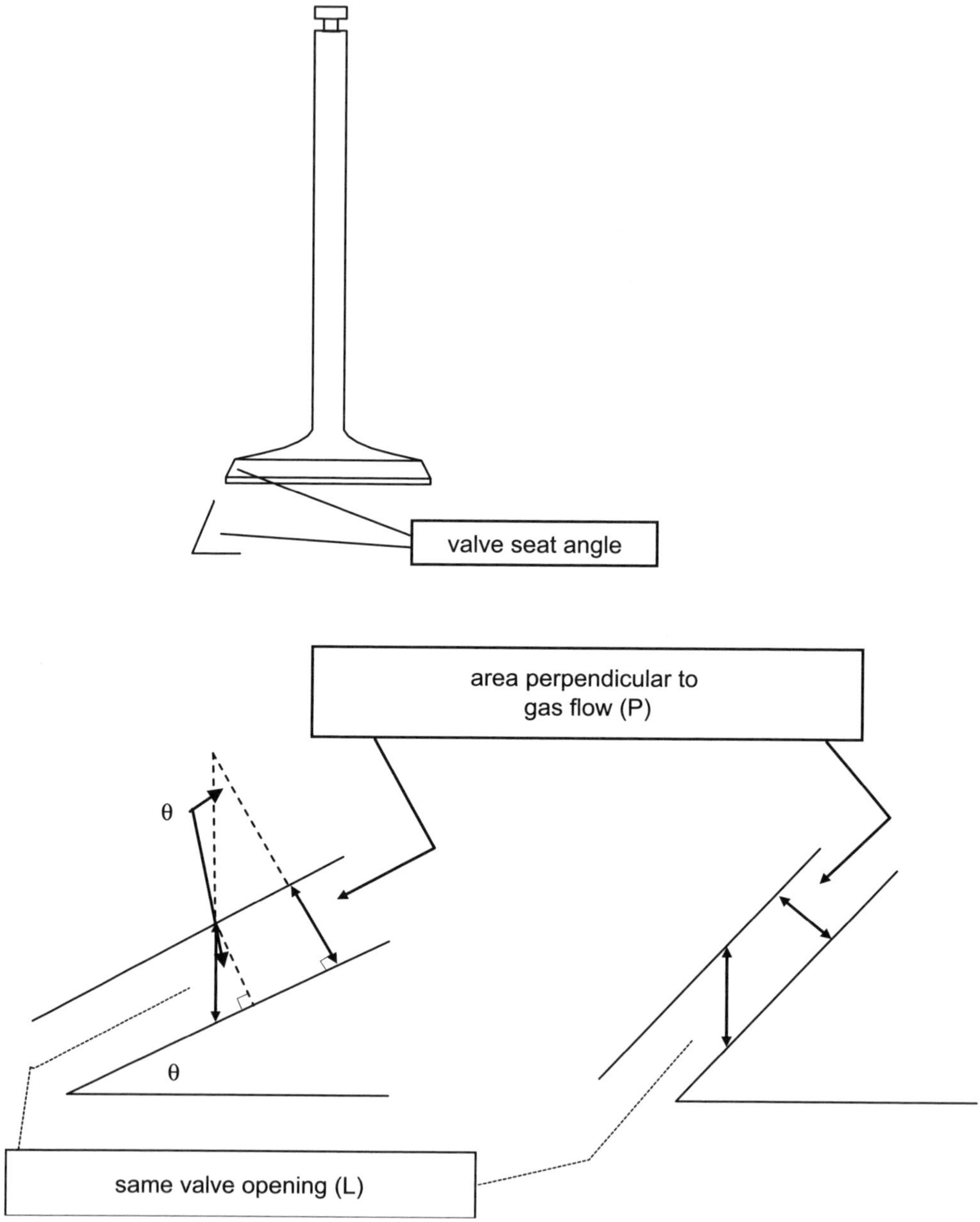

Fig. 13.9. Valve angles

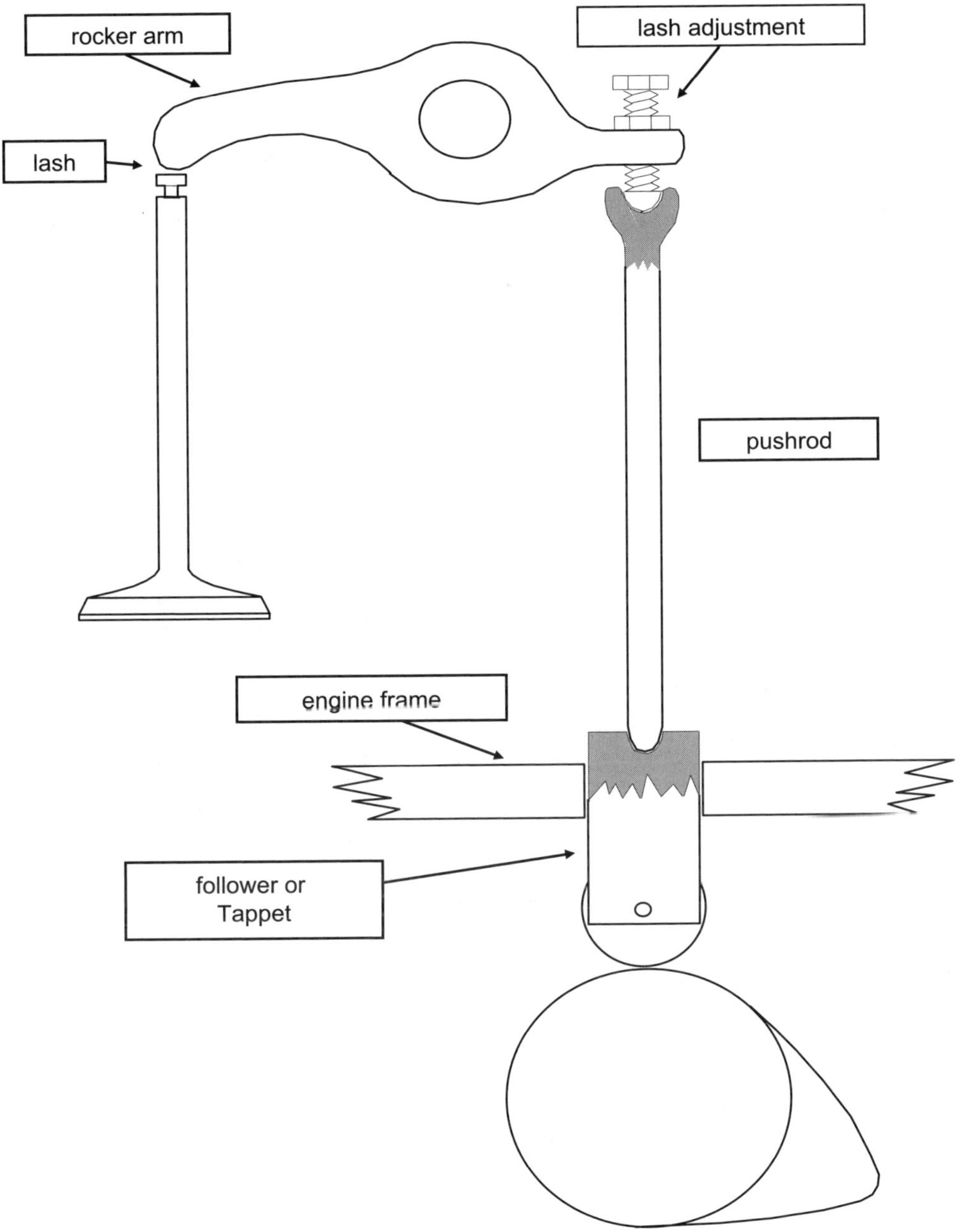

Fig. 13.10. Valve train

lash adjuster and puts some lash back in. This movement is very small for an individual valve opening and closing. The piston, however, will be farther up in the lash adjuster cylinder when the engine is up to operating temperature. This phenomenon is depicted in fig. 13.11.

Slow-speed engines formerly used pushrods, but now use a hydraulic actuator with an air spring, as shown in a fig. 13.12.

Today the trend is toward a completely computer-controlled, slow-speed engine. The fuel injection, exhaust valve operation, air-starting operation, and cylinder lubricator operation are controlled by computer-controlled solenoid valves. This eliminates the camshaft. The camshaftless engine can time the opening of the exhaust valve at slow speeds so exhaust temperatures do not drop below 150 °C. This eliminates condensation in the stack. Currently this system is limited to slow-speed engines because of the speed of the solenoid valves.

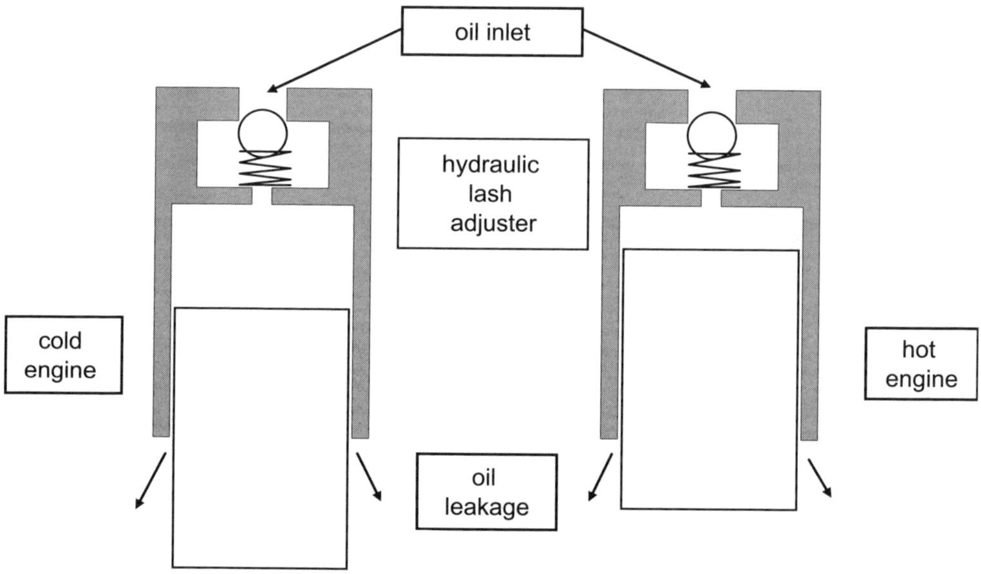

Fig. 13.11. Hydraulic lash adjuster

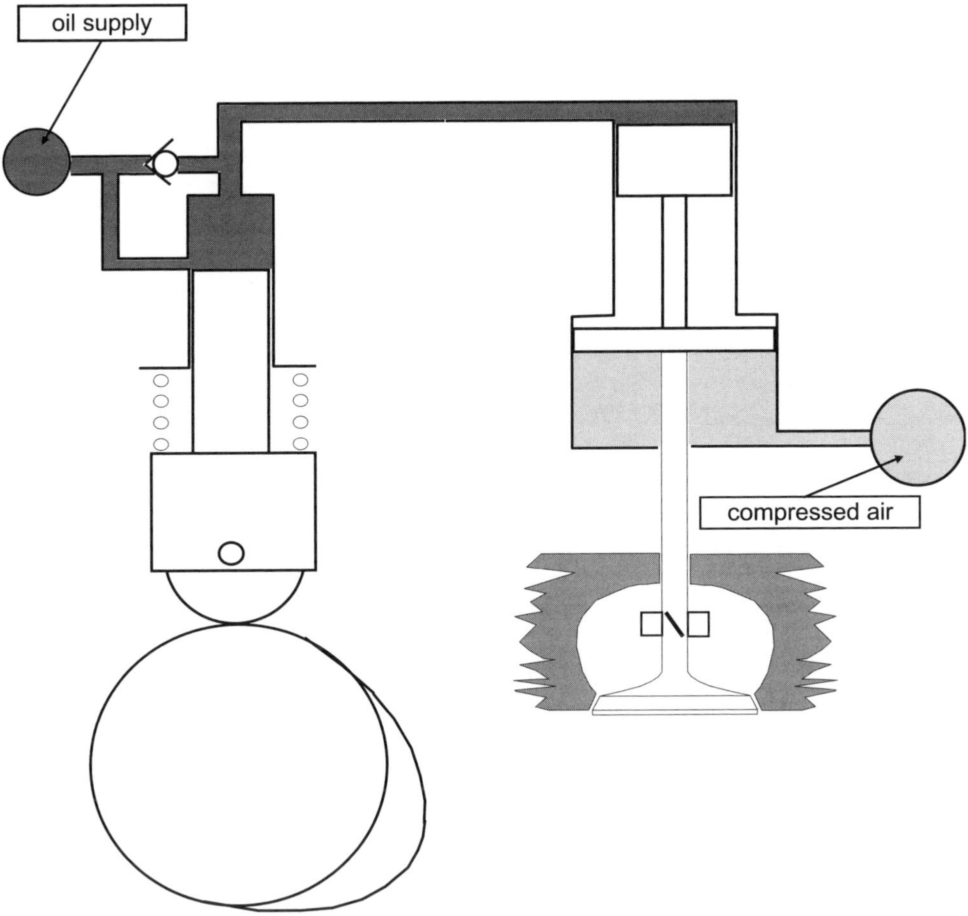

Fig. 13.12. Simple diagrammatic view of a hydraulically operated exhaust valve for a slow-speed engine

Study Questions

1. Which engine does not have an air start valve?
   A. slow speed
   B. medium speed
   C. high speed
   D. Crosshead type
2. All modern slow-speed engines have this type of cylinder liner:
   A. wet
   B. dry
   C. differential
   D. integral
3. Where is the carbon scraper located in diagram A?
   A. 1
   B. 2
   C. 3
   D. 4
4. What can be reconditioned on an intake valve?
   A. stem
   B. head
   C. margin
   D. seat
5. The object in diagram A is held closed by:
   A. hydraulic pressure
   B. pushrod
   C. spring
   D. cam

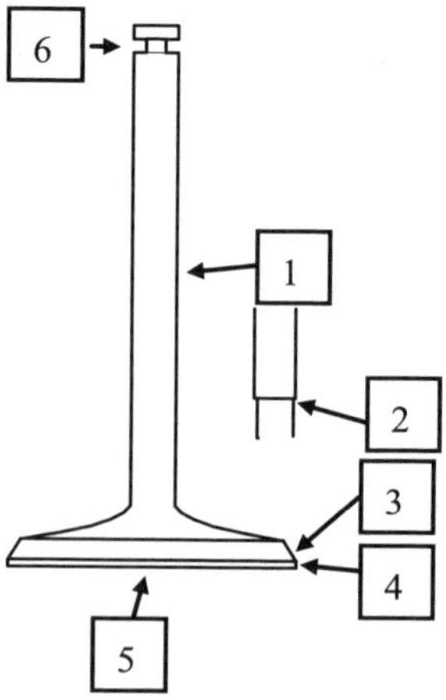

Diagram A

6. Where would Stellite be located in diagram A?
   A. 1
   B. 2
   C. 3
   D. 4
7. Where is the seat located in diagram A?
   A. 1
   B. 2
   C. 3
   D. 4
8. What characteristic of a valve spring influences whether cam follower bounce will occur?
   A. fatigue strength
   B. elastic limit
   C. proportional limit
   D. spring constant
9. Which of the following is used to aid in cooling the exhaust valve seats in a slow-speed engine?
   A. water-cooled valve seat insert
   B. valve rotator
   C. sodium-filled valve
   D. A and B
10. The smaller the valve seat angle:
   A. the smaller the vertical movement of the valve
   B. the larger the closing force
   C. the larger the gas flow area
   D. the smaller the gas flow area
11. Which cylinder liner has a water jacket as part of its construction?
   A. wet
   B. dry
   C. ported
   D. integral
12. Stellite is found on what valve?
   A. intake
   B. exhaust
   C. air start
   D. indicator cock
13. Where would the split keeper be located in diagram A?
   A. 6
   B. 1
   C. 2
   D. 3

14. Wire drawing can be detected by looking at the:
    A. cooling water temperature
    B. inlet air temperature
    C. exhaust pyrometer
    D. viscometer
15. Hydraulic valve lifters extend by using:
    A. lube oil pressure
    B. fuel oil pressure
    C. air pressure
    D. spring force
16. What allows for the thermal expansion of the valve gear?
    A. lash
    B. temperature controller
    C. pyrometer
    D. rotocap

Chapter 14
Fuels and Lubricants

Fuels

Number	Distillate	Cetane No
	Content	
1	High-grade Distillate	40-65
2	All Distillate	40-65

The above fuels are 100 percent distillate and used in high-speed engines. The fuels below are used in medium- and slow-speed engines.

3	No. 3 fuel exists at present.	
4	Distillate and Residual Mixture	30-45
No. 4 fuel requires no heating. All fuels beyond this require heating.		
5 Bunker B	Distillate with more Residual in the mixture	
6 Bunker C	All Residual	

Bunker C is mainly residual oil, but to pump this oil a certain minimum viscosity must be maintained, and some recycled solvents may be mixed in. This is to allow the heating and pumping of the fuel after it is stored without heat in double bottoms or deep tanks. It is possible to have residual fuel that cannot be heated and pumped out of a tank after it has cooled to ambient temperature. If the fuel is viscous enough, the heating coil in the fuel tank can reduce the viscosity around the coil where the pump suction is located and the transfer pump pumps this oil. This leaves the rest of the fuel as a solid mass in the tank. Some specialty companies offer portable heaters and mixers to remedy this situation. Obviously, it is better not to buy this type of fuel.

Fuel Oil Parameters

The Cetane Number is the percent of cetane mixed with alphamethylnapthaline, which acts like the fuel is being tested in a diesel engine. Cetane is a hydrocarbon molecule that has the shortest ignition delay, alphamethylnapthaline the longest.

Specific gravity is the density of fuel divided by the density of water, both measured at 60 °F. This parameter is used to convert from volume to weight since the density of water is universally known.

API Gravity is a version of specific gravity used by the American Petroleum Institute (API).

$$\text{DegreesAPI} = \frac{141.5}{\text{sp.grav.}} - 131.5$$

This means that water has an API gravity of ten. This parameter is useful in using charts published by the API for such items as heating value and densities.

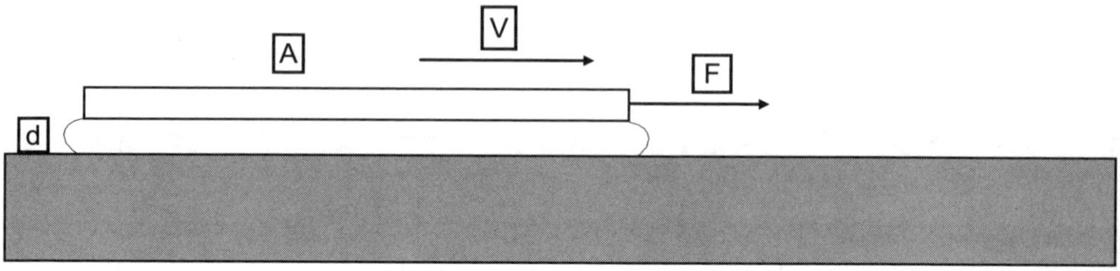

V velocity of plate
d thickness of fluid
F force to move plate at plate velocity
A area of plate

Fig. 14.1. Viscosity—resistance to flow

Dynamic Viscosity or Absolute Viscosity == $\dfrac{F/A}{V/d} = \mu \quad \dfrac{lb \bullet \sec}{ft^2}$ or $\dfrac{dyne \bullet \sec}{cm^2}$ or $\dfrac{Sl}{ft \bullet \sec}$ or $\dfrac{g}{cm \bullet \sec}$

Using W=MG thus $lb = \dfrac{Sl \bullet ft}{\sec^2}$ or $dyne = \dfrac{g \bullet cm}{\sec^2}$

In the metric system, the unit of measurement is called the poise or $\dfrac{dyne \bullet \sec}{cm^2}$ $\dfrac{g}{cm \bullet \sec}$

The Kinematic Viscosity $\nu = \dfrac{\mu}{\rho}$

The absolute viscosity and the density must be at the same temperature.

$$\nu = \frac{\mu}{\rho} \qquad \frac{ft^2}{\sec} = \frac{Sl}{ft \bullet \sec}\,\frac{ft^3}{Sl} \qquad \frac{cm^2}{\sec} = \frac{g}{cm \bullet \sec}\,\frac{cm^3}{g}$$

In the metric system, the unit of kinematic viscosity is the Stoke. $\dfrac{cm^2}{\sec}$ The usual unit, however, is the centistoke (cSt). $\dfrac{mm^2}{\sec}$

It is important to use the kinematic viscosity when working with oil, since the absolute viscosity and density change substantially with temperature.

This idea of viscosity has a mathematical use in the Reynolds Number and other fluid equations. The trend today is to use cSt to name the grades of fuel. Heavy fuel oil is designated as (HFO-viscosity in cSt) such as HFO-600. Some blends of heavy oil and distillate are called intermediate fuel oil (IFO-viscosity in cSt), such as IFO-240. Viscosity in the marine field also is measured in seconds. This is done by having a certain size pot with a specified diameter hole in the bottom with the oil at a specified temperature. The time it takes for the pot to empty, measured in seconds, is the viscosity.

The usual names are as follows:

1. Redwood (RW1 or RW2)
2. Saybolt Seconds Universal (SSU)
3. Saybolt Seconds Furol (SSF)
4. Engler

An example of these values for Bunker C and No. 2 diesel oil are:

	API	RW1	SSU	cSt
	Grav	100 ° F	100 ° F	122 °F
				50 °C
No. 2 Diesel	32	40	45	5.5
Bunker C	15	3500	4000	380

The viscosity usually required for injection is about 50 to 100 Saybolt Seconds Universal. The No. 2 Diesel Fuel is already there. That is why it is used on high-speed engines that have no fuel oil heating system. The Bunker C oil has to be heated to more than 250 °F to achieve this viscosity. MAN B&W require the fuel viscosity to be ten to twelve cSt at the fuel injection pumps.

Calorific Value or Heating Value

Calorific value or heating value is the heat given off by the burning of a unit mass of fuel. This is usually measured in Btu/lb in the English system. A higher heating value occurs if the water in the products of combustion is condensed. The lower heating value is the value of the heat given off if the water vapor is not condensed. The water is not condensed when burning fuel in a diesel engine.

Diesel Index

This is an empirical relationship derived from the API gravity and the mid-boiling point temperature. It is very close in value to the cetane number. The mid-boiling point temperature is the temperature of a pot of fuel in degrees Fahrenheit when 50 percent (by volume) of the fuel is boiled off.

Volatility

This is the temperature in degrees Fahrenheit when 10 percent (by volume) of the fuel is boiled off and collected.

Figure 14.2 represents the typical rise in temperature as a fuel is heated and distilled. The lower the temperature that the fuel starts to boil off, the more volatile the fuel.

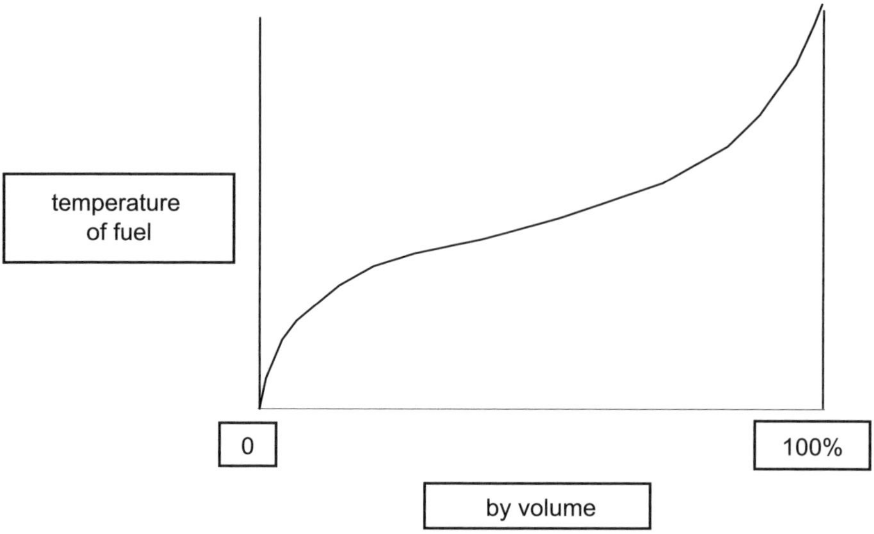

Fig. 14.2. Volatility

Flash Point

Flash point is the temperature in degrees Fahrenheit that the fuel will flash, but not continue to burn if a flame is passed over a cup of the fuel. A closed cup test is used on the more volatile fuels and the opened cup test is used on the less volatile. This test has little to do with the performance of the fuel in an engine, but shows how dangerous the fuel can be in handling and storage. The minimum flash point for No.1 and No. 2 diesel fuel is 110 °F, as required by the USCG.

Fire Point

Fire point is the temperature in degrees Fahrenheit that the fuel will continue to burn if a flame is passed over it. This temperature is usually twenty to fifty degrees above the flash point.

Cloud Point

Cloud point is the temperature at which wax will start to precipitate out and plug filters. This indicates whether you will have a problem getting distillate fuel to your engine during cold weather.

Pour Point

Pour point is the lowest temperature at which the fuel can be pumped. This point indicates whether the fuel can be reheated and pumped after it is allowed to cool to ambient temperatures. A minimum pour point usually is specified for a fuel.

Carbon Residue

Carbon residue is the percent by weight of carbon residue left in a pot after it is heated but not burned. The most popular apparatus today is the Conradson Carbon Test Apparatus. When using this particular apparatus the percent is called the Conradson Carbon Number. This number gives an indication of a fuel's tendency to form carbon in an engine.

Sulfur Content

Sulfur content is the percent of sulfur by weight in the fuel. The sulfur forms sulfur dioxide and sulfur trioxide when burned. These compounds dissolve in water and form sulfuric acid. The sulfur content is limited by international and local restrictions.

Ash Content

Ash content is the percent of material left after the fuel is heated and burned at high temperatures. These include silicates, oxides of aluminum (cat fines), vanadium, sodium and other minerals. These promote wear in the injection equipment, pistons, piston rings, and liners. They also increase engine deposits. Liquid heavy fuel has an ash content in the tenths of a percent, while coal has an ash content in the percents. This is over an order of magnitude higher. Thus, burning a coal slurry fuel can cause great wear problems. This slurry is made from powdered coal and fuel oil. This alternative fuel probably will become more important because of the price of fuel oil. The ash can be electrostatically separated from the coal when it is in powdered form.

Sediment Content

Fuel is mixed 50/50 with benzol and centrifuged. The amount of sediment collected is expressed as a percent by volume. This indicates how fast filters are plugged and how often purifies shoot. This sediment contains ash and large molecules of carbon that do not dissolve in benzol.

Water Content

Fuel is centrifuged and the amount of water is expressed as a percent by volume. Since water has a higher surface tension then fuel, if the water content is too high it may retard the flow through the fuel injector holes enough to blow off the tip. A highly emulsified fuel and water mixture actually is good. The explosive boiling of the water upon injection aids the atomization. It also reduces the oxides of nitrogen produced. This fuel treatment requires special equipment to emulsify the fuel and water (homogenizer). Just shaking some fuel with water will not work.

Vanadium and Sodium Content

The amount of vanadium and sodium content is expressed in parts per million. These elements form a eutectic solid solution. Another example of this is antifreeze and water. Though not a solid solution, the effect is the same. Each constituent has a higher melting point than the mixture. The sodium and vanadium eutectic lower the melting point of the sodium and vanadium salts, allowing them to stick to the exhaust valve seats and form exhaust valve deposits. If the melting point were higher, they would not melt, but remain solid and bounce off the valve seat. Sodium and vanadium salts also can combine with the piston crown to lower its melting point and promote burn down of the piston. Vanadium is usually the only component specified by the engine manufacturer. Sodium is always in the oil because it comes from the ground.

Vanadium pentoxide and iron oxide act as a catalyst for the formation of sulfur trioxide. This combines with water to form sulfuric acid. Water is one of the products of combustion in burning a hydrocarbon. The sulfur trioxide also raises the dewpoint for sulfuric acid. This chain reaction not only produces sulfuric acid, but promotes its condensation.

Additives

Distillate fuel has a few additives:

1. Amyl nitrate can be used as a cetane improver, but is expensive and usually not used.
2. Detergents and solvents are available to help keep injectors clean.
3. Biocides are used to keep bacteria from growing, contaminating the fuel, and plugging up fuel filters.

Magnesium has been proposed as an additive to improve heavy fuel oil. Magnesium can raise the melting point of vanadium and sodium salts, but this is expensive and not widely used at present.

Power Research Inc. has a product they claim can reduce oxides of nitrogen 21 percent and sulfur dioxide by 10 percent, double exhaust valve life, and keep turbocharger blades, waste heat boiler, and tubes soot free. However, they do not explain what is in their product.

Wear, Abrasion, and Corrosion

These processes affect the cylinder liner, piston, and piston rings.

Abrasion particles are necessary for this to occur. The particles can come from the following sources:

1. Carbonaceous particles. These particles come from the incomplete combustion of the fuel, which tends to occur with faulty injection. These particles are especially hard if they contain sulfur. Most heavy fuel oil does contain sulfur.
2. Ash. Residual fuel oils contain a small, but significant, amount of ash.
3. Particles that come in with the air needed for combustion.

Corrosion

The main source of corrosion when burning heavy fuel is the sulfur contained in the fuel. Sulfur burns to form sulfur dioxide. Sulfur trioxide also is formed, especially in the presence of vanadium pentoxide and iron oxide. The sulfur trioxide forms sulfuric acid when dissolved in water. One of the products of combustion of a hydrocarbon is water.

Sulfuric acid attacks the cylinder liners, pistons, and piston rings, and the lube oil film on the wearing surfaces of these components.

Abrasion and corrosion work together to accelerate wear. Corrosion changes the surfaces chemically into a weaker material and abrasion knocks off this weaker material, allowing more corrosion to occur.

The faulty injection that causes the carbon particles also can hit the cylinder liner walls, washing away the lube oil film and accelerating wear.

Proper cylinder wall temperatures help prevent corrosion. If the cylinder wall temperature is high enough, moisture with sulfuric acid does not condense. This can be done by recirculating hot water, instead of restricting water flow for temperature regulation. All modern diesel engines use this practice. The liner cooling water temperature is kept to a temperature difference between inlet and outlet of about 20 °F. This gives the liner the proper temperature throughout its length. With too high a temperature difference, the lower part of the liner is too cool. The camshaftless engine keeps the cylinder wall temperature above 150 °C during low loads by varying exhaust valve timing.

Piston Deposits

Top carbon deposits are made up of carbonaceous-asphaltic matter, ash, sulfur compounds, and inlet air particles. These materials are produced mainly by the fuel, but some are from the lube oil. If these deposits buildup enough to get into the clearance between the piston and the cylinder liner, they can also get between the compression rings and ring grooves. Deposits in these areas cause compression rings to stick. This produces a lack of compression and poor combustion, which creates more deposits. At the same time, the increased blow by contaminates the crankcase lube oil in trunk piston-type engines. The crosshead engine does have this problem because of the piston rod seal separates the crankcase from the underside of the piston.

Some deposits may also settle on the piston skirt, but these usually are from lube oil. The three basic ways to fix the deposit problem are:

1. Burn distillate fuel. For most main propulsion engines this is not an economical choice.
2. Optimize the fuel injection system. Make sure all components are within specifications. The present trend is to use very high injection pressures (5,000 psi or more) to ensure the complete combustion of heavy fuel oil.
3. Modern lubricating oils. These oils must be clean burning, have low ash content, and contain an additive package that deals with the problems of burning heavy fuel oil. It may seem strange that lube oil must be clean burning, but a significant amount burns in the upper reaches of the combustion chamber.

Lubricating Oil Additives

One of the main reasons that heavy fuel oil containing sulfur can be burned in modern propulsion engines is that modern lubricating oil contains an additive package.

Oxidation Inhibitors

All lubricating oil oxidizes if the temperature is high enough and the surface area great enough. Oxidized lubricating oil is no longer oil and usually is acid. This promotes corrosion and poor lubrication, so oxidation inhibitors are essential. The temperature limit of the top ring in medium-speed engines is 200 °C or 392 °F to make sure the lubrication oil does not oxidize.

Detergent Additives

Detergent additives help keep deposits finely suspended in the lube oil so they pass through the engine without harming it. This suspended matter can be removed by the lubricating oil purifier and filters. The detergent additives are especially useful in keeping the piston rings clean and free.

Alkaline Additives

Alkaline additives are used to neutralize the acid products produced by burning the sulfur in heavy fuel oil and acid products produced by the oxidation of the lubricating oil. If this acid is not neutralized, it attacks crankshaft bearing material and all the combustion chamber surfaces. Too high an alkaline additive content can cause problems. Too caustic an oil attacks the same components just as too much acid.

Crosshead engines have a separate cylinder lubrication system. This system uses oil that is more alkaline and has a higher viscosity than the crankcase oil. The cylinder lubrication system is a once through system that supplies cylinder oil with small positive displacement piston pumps run from the camshaft. The pumps supply the oil to lubricating quills screwed into the cylinder liner wall. These quills are equipped with a spring-loaded piston and a check valve. Oil is supplied to the cylinder wall during periods of low cylinder pressure. If the pump is pumping during high cylinder pressures, the spring-loaded piston stores the oil until the pressure drops and the oil moves through the check valve.

The piston rod seal on crosshead type engines essentially keeps the products of combustion out of the crankcase. This allows a more alkaline oil to be used in an area around the piston that has no bearings. The viscosity of the oil used is higher, since this area is running at a higher working temperature than the crankcase.

Trunk piston engines have to use a compromise oil that is less alkaline since this oil lubricates the pistons and the crankcase. Improved gas sealing with better piston ring packages also help the trunk piston engine burn heavy oil. Crankcase temperatures usually are kept higher in medium-speed engines burning heavy oil to avoid condensation of any moisture which can contain sulfuric acid.

System Lubricating Oil Testing

If the lubricating oil purification systems are working properly and no fuel or water leaks into the crankcase, the crankcase oil of medium- and slow-speed engines is never completely changed at regular intervals. A significant amount of makeup oil is added as lubricating oil and is burned in the upper reaches of the combustion chamber. In medium- and slow-speed engines the lubricating oil in the crankcase is changed continuously. Because of this, it is important to test the lubricating oil in the crankcase of main propulsion engines.

System lubricating oils undergo various tests to ascertain their condition.

Neutralization Value is the number of milligrams of potassium hydroxide that will neutralize the acid in one gram of oil. Since the products of oxidation of lubricating oil and the sulfur products of combustion are acidic, this test is an indication of how fast the oil is oxidizing. However, this test is outdated since most lubricating oil used in heavy fuel burning engines has alkaline additives.

Total Base Number (TBN) is the amount of potassium hydroxide equivalents in one gram of oil. This indicates how much acid the oil can neutralize. This is a better test for the lubricating oil in heavy fuel burning engines. A decrease in this number indicates how bad the contamination is from the combustion products, specifically from burning sulfur and partially from oil oxidation products.

MAN-B&W recommends for crosshead engines:

Fuel Burned	System Oil	Cylinder Oil
No. 2 Diesel	SAE30 TBN 2-10	SAE40 TBN 2-10
3600 RW1	SAE30 TBN 2-10	SAE40 TBN 16-25

The viscosity has to be within specifications. Too viscous an oil uses up horsepower. Too thin an oil increases wear.

Medium-speed trunk piston engines usually add makeup oil that is more alkaline than the system oil. This allows the system oil to reach equilibrium as the alkaline additives neutralize the sulfuric acid produced during combustion.

Viscosity. The viscosity of lubricating oil is increased by the oxidation products of the oil and carbon particles being produced by improper combustion. Improper combustion can be caused by faulty injection equipment, insufficient air, and improper fuel temperatures. A 10 percent increase in the viscosity of the lube oil indicates a serious problem and the oil may have to be renewed and the problem found. The viscosity of the lubricating oil may be reduced by dilution with light fuel oil. If heavy fuel oil dilutes the lubricating oil, the viscosity may go up or down, depending on the viscosity of the heavy oil.

Flash Point. This indicates fuel oil dilution, regardless of the type of fuel. The flash point of most lubricating oils is 420 °F. The flash point of the heaviest fuel oil is about 200 °F.

Water Content. This test usually exposes any cooling water leaks. A water content of .2 percent is permissible. Anything over this aids sulfuric acid and the acid oxidation oil products in attacking engine internals.

Insoluble Matter. These are the solids insoluble in heptane and benzene after centrifuging. They can be any of the following:

- atmosphere dust
- carbon or soot from incomplete combustion
- metallic wear products from internal engine parts

Analysis of these particles can show if the air filters are failing or what part of the engine is wearing abnormally. This can be done only if the particles can be identified and counted. To do this, the particles are separated magnetically and by mass. Separation by mass is called mass spectrographic analysis. After the particles are separated, they can be analyzed through a microscope using natural light, polarized light, and X-Rays.

Companies that perform this type analysis keep a database of all the tests on different makes of engines. They can tell when something is abnormal and exactly what the abnormality is.

The lubricating oil for this test must be taken from an unfiltered source on the engine.

Filters

The three critical times in an engine's life where extra filtration is needed are:

1. When the engine is first run on the testbed.
2. When the engine is installed in the ship and run for the first time with the ship lubricating oil system.
3. Anytime the engine or ship lubricating system has major work done.
4. n these situations, foreign matter can get into the lubricating system and lubricating passages inside the engine. Some shipping companies have skid-mounted lubricating oil systems that can be brought to the shipyard to provide extra filtration during these situations.

All filtering systems are a compromise. The smaller the size of the particle removed, the bigger the filter has to be to keep the pressure drop across the filter to reasonable levels. Thus, bypass lubricating oil filters are finer filters than full-flow filters. Because not all the oil goes through the bypass filter, it can be a small fine filter. In this way, costs can be kept down.

Metal Gauze Filter. These are filtering screens that can be manufactured to any size. Coarse screens are used as inlet strainers for pumps. Fine screens that filter particles down to five microns are used in self-indexing, back-flushing filters in the fuel oil service system. (A micron is a millionth of a meter.)

Metal Edge. These filters are comprised of metal or plastic disks separated by shims. They can filter out particles down to .05mm (.002 ″). The flow is from the outside in, so the filter can be cleaned by back flushing. They also are commonly cleaned by a set of rotating mechanical fingers that travel around the spaces between the disks, but the particles end up in the bottom of the housing, and the filter must be taken apart to remove this sludge.

Special Edge Type Filter. This filter uses plastic washers with triangular channels cut across the flat surface.

The flow is from the outside in. The smaller part of the groove is on the outside. This is to ensure particles smaller than specified cannot jam the groove. This filter can be cleaned by back flushing or disassembly. This filter can remove particles down to five microns.

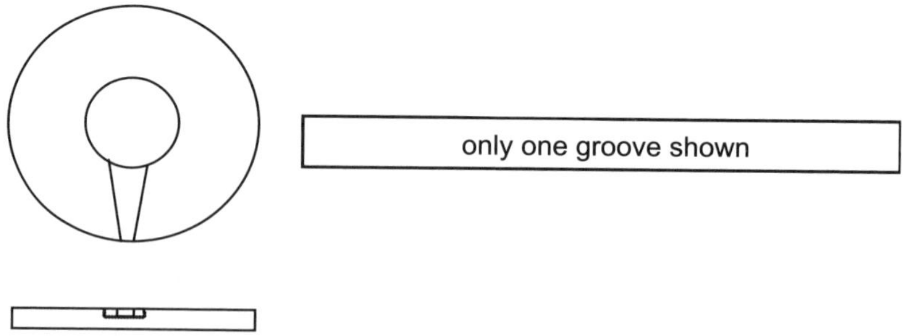

Fig. 14.3. Special edge filter element

5 microns   .1mm $5 \times 10^{-6} m \rightarrow 5 \times 10^{-3} mm \rightarrow 5 \times 10^{-2}$

.1mm = .004" Thus 5 microns = .0002"

Surface Filters. These filters use elements of paper or felt. Flow is from the outside in and the elements cannot be cleaned, they must be replaced. These filters can remove particles down to two microns. Most of these disposable filtering elements have a bypass valve built into the housing, so it is important to change the element before it becomes plugged. If not, you may think the filter is working, but it has been bypassed. Some housings have an indicator to tell when this happens. Large filtering systems have pressure gauges around the filter to indicate when the element is plugged.

All surface filters work as absorbent-type filters, which means they swallow impurities like a sponge. They can remove very fine particles and water until the element becomes saturated.

Chemically treated paper and Fullers Earth (Diatomacious earth) filters work on the principal of adhesion. This means a thin layer of molecules of an impurity can adhere to the surface of the solid filtering element. This filter can take out gum, tar, and acids that are dissolved in the oil. Specifications for this type of filter must be chosen carefully because this filter can also remove the additives in the oil.

Study Questions

1. A fuel with a short ignition delay will have a:
   A. high octane number
   B. low cetane number
   C. high cetane number
   D. high TBN number
2. What viscosity takes into account the density of the fluid?
   A. absolute viscosity
   B. kinematic viscosity
   C. Redwood viscosity
   D. Saybolt viscosity
3. What parameter indicates that fuel oil can be pumped at low temperature?
   A. gel point
   B. pour point
   C. carbon residue
   D. ash content
4. 4. Which fuel oil parameter indicates if your filters will plug at low temperature?
   A. cloud point
   B. fire point
   C. carbon residue
   D. ash content
5. Which fuel oil parameter indicates how much corrosion will occur on the cylinder liner?
   A. carbon residue
   B. sulfur content
   C. ash content
   D. sediment content
6. What is the specific gravity of water?
   A. .1
   B. 62.4
   C. 1
   D. 10
7. 7. In the named IF-380 fuel, 380 stands for the:
   A. viscosity measured in Centistokes
   B. viscosity measured in Centipoise
   C. viscosity measured in seconds Redwood
   D. specific gravity in degrees API
8. Which heating value includes the heat of vaporization of water?
   A. TBN
   B. BN
   C. LHV
   D. HHV

9. Adding magnesium to heavy fuel oil would:
   A. raise the cetane number
   B. lower the cetane number
   C. raise the melting temperature of sodium and vanadium salts
   D. lower the melting temperature of sodium and vanadium salts
10. When is the crankcase oil changed in a main propulsion engine?
   A. continuously
   B. never
   C. monthly
   D. every 200 hours
11. The more hours crankcase oil has been used the:
   A. lower the viscosity
   B. lower the flashpoint
   C. higher the flashpoint
   D. higher the viscosity
12. The self-indexing back-flushing filters used for final filtering of lube oil and fuel oil use what type of filtering medium?
   A. absorbent
   B. adsorbent
   C. metal gauze
   D. metal edge
13. Which engine is least affected by combustion products contaminating the crankcase?
   A. slow speed
   B. medium speed
   C. high speed
   D. opposed piston
14. Which oil has the highest viscosity?
   A. camshaft oil
   B. governor oil
   C. crankcase oil
   D. cylinder oil
15. The TBN test of lubricating oil is for what property?
   A. acid neutralizing
   B. alkalinity neutralization
   C. oxidation neutralization
   D. detergent neutralization
16. What lube oil test can lead to pinpoint specific component failure problems in a diesel engine?
   A. viscosity
   B. water content
   C. insoluble matter
   D. sulfur content

17. A micron is:
    A. millionth of a millimeter
    B. millionth of a meter
    C. millionth of an inch
    D. millionth of a mil

Chapter 15

Basic Construction (Static Parts)

Engine Construction

The two basic types of engine construction are block construction and bedplate construction. We will first look at the bedplate construction. This construction is mostly used on slow-speed, crosshead engines. The bedplate is the foundation of the engine. The A-frame is bolted on top of the bedplate. The cylinder block is then bolted on the A-frame. The cylinder liner is put into the cylinder block and the head is bolted to the cylinder block with head bolts or studs and nuts.

The tie rods hold the cylinder block, A-frame and bedplate together against the force of the firing pressure. Only one set of tie rods are shown fig. 15.1. A set is between each cylinder and at each end.

The bedplate is made up of welded steel plates and steel castings. All flat areas use steel plates, while the bearing pockets are castings. The bedplate is welded, using full penetration, down hand welding. It is made in one piece for the engine and contains cross members that hold the main bearings of the crankshaft. The top is open to allow the crank throw to rotate into the A-frame. The sides are usually double walled with flanges top and bottom to produce a double web I-beam. This will make the bedplate a strong box girder. The corners are reinforced with gussets to prevent the box from racking.

Fig. 15.2 represents a simplified version of one bay of a welded bedplate.

The bottom of the bedplate is covered with a thin piece of steel to collect lubricating oil. A pipe connects the bedplate bottom to the lubricating oil sump located in the double bottom below the engine.

The A-frame, or columns, is bolted onto the bedplate. The A-frame is made of welded steel plates. It may be made in one piece or for a large twelve-cylinder engine, in two pieces. In any case, it will encompass many bays of the crankshaft. The A-frame will contain the guides for the slippers of the crosshead bearing, tubes for the tie rods, and thrust bolts to hold down the main bearing cap.

Hydraulic thrust bolts incorporate a hydraulic piston, hydraulic cylinder and a locking nut to hold down the bearing cap. The cylinder is pressurized and then the nut is run under the cylinder to keep the force on the bearing cap when the pressure is released. This eliminates the projections most bearing caps have to contain the bearing hold down bolts.

The bedplate and the A-frame are welded, using full penetration down hand welding. Set procedures are used in the welding sequences to eliminate warping of these components. Warping components made up of welded plates is always a danger. This is especially true

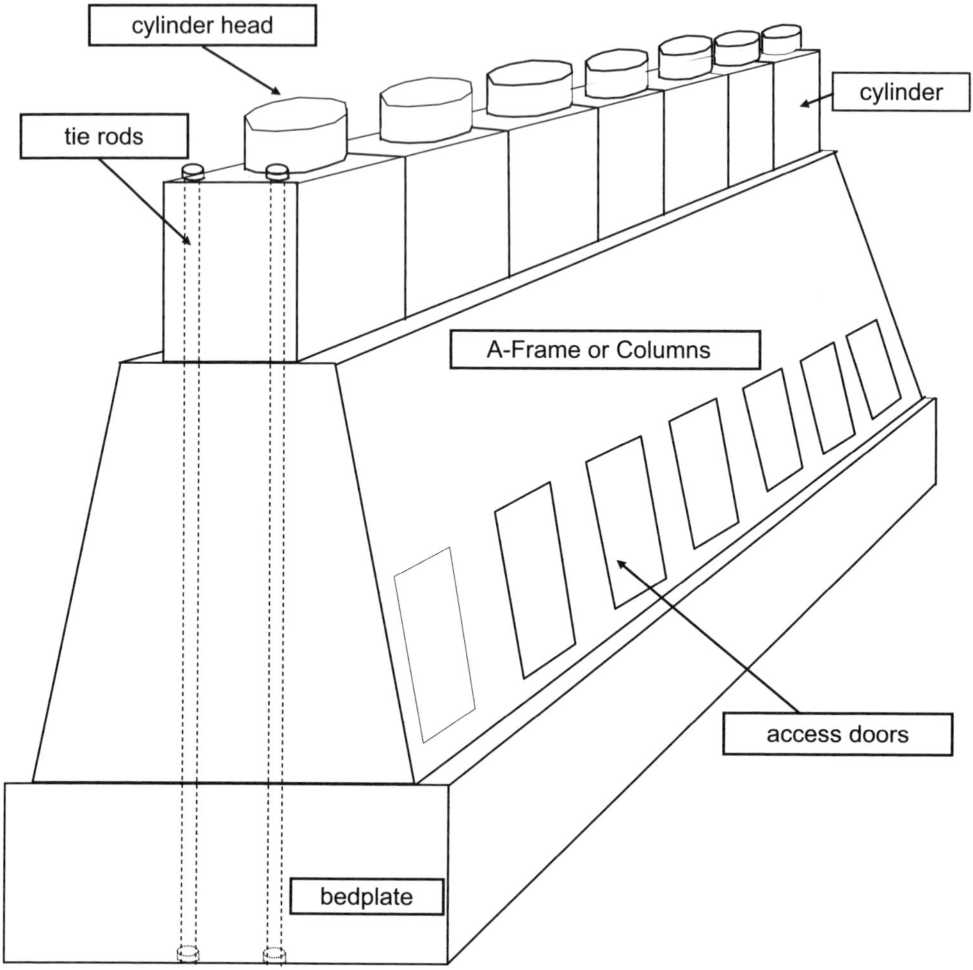

Fig. 15.1. Bedplate construction

for such large components as the bedplate and A-frame. After the welding is completed, the component is put in an annealing oven. This is done to relieve the built-up stresses in these components due to the solidification and cooling of the weld material. After annealing, the final machining of critical dimensional surfaces is performed.

Fillet welding is done when the joining parts are not highly stressed. This method is cheap and less time consuming. The two pieces of metal just have to be butted against each other and welded. However, this will produce a stress riser. A crack can form in the weld metal where it cannot be seen.

The full penetration down hand welding eliminates the stress riser problem by completely filling in the joint with weld material. As shown in fig. 15.4, the material must be

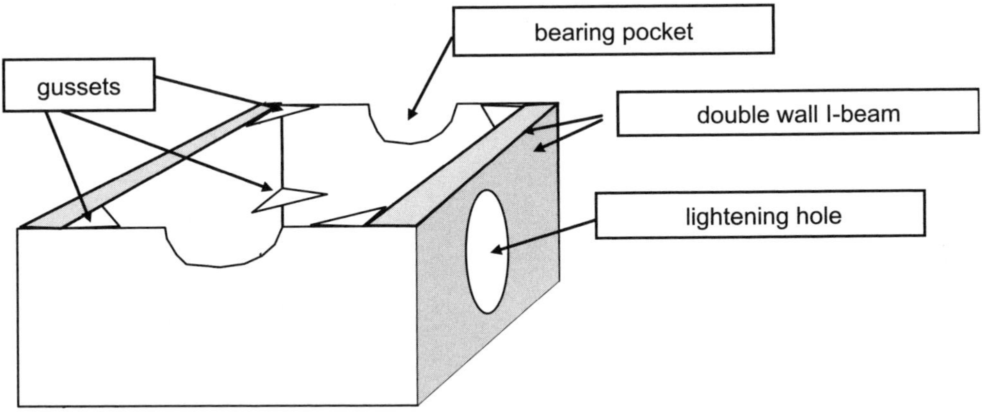

Fig. 15.2. One bay of bedplate

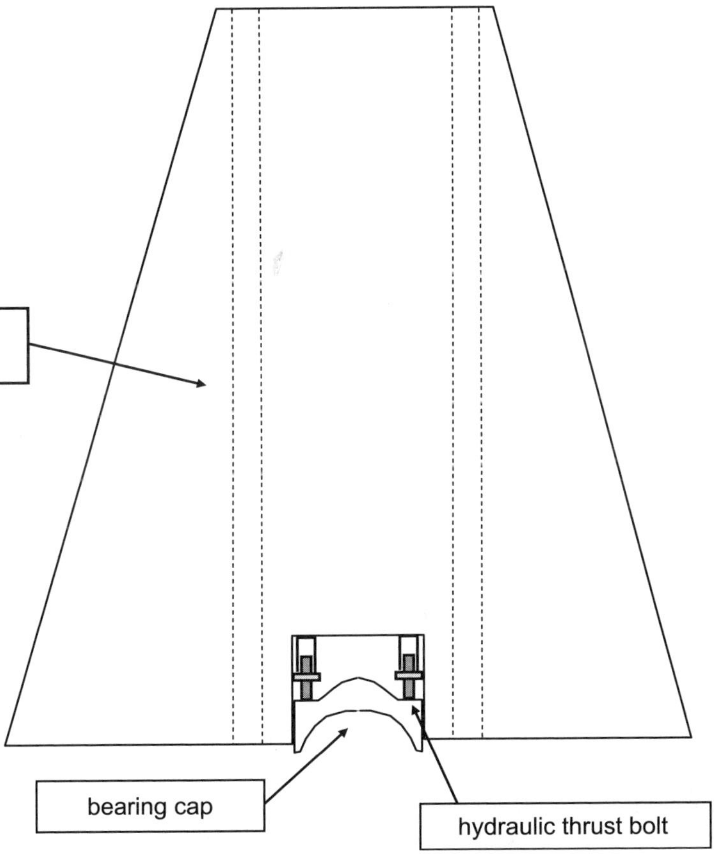

Fig. 15.3. A-frame

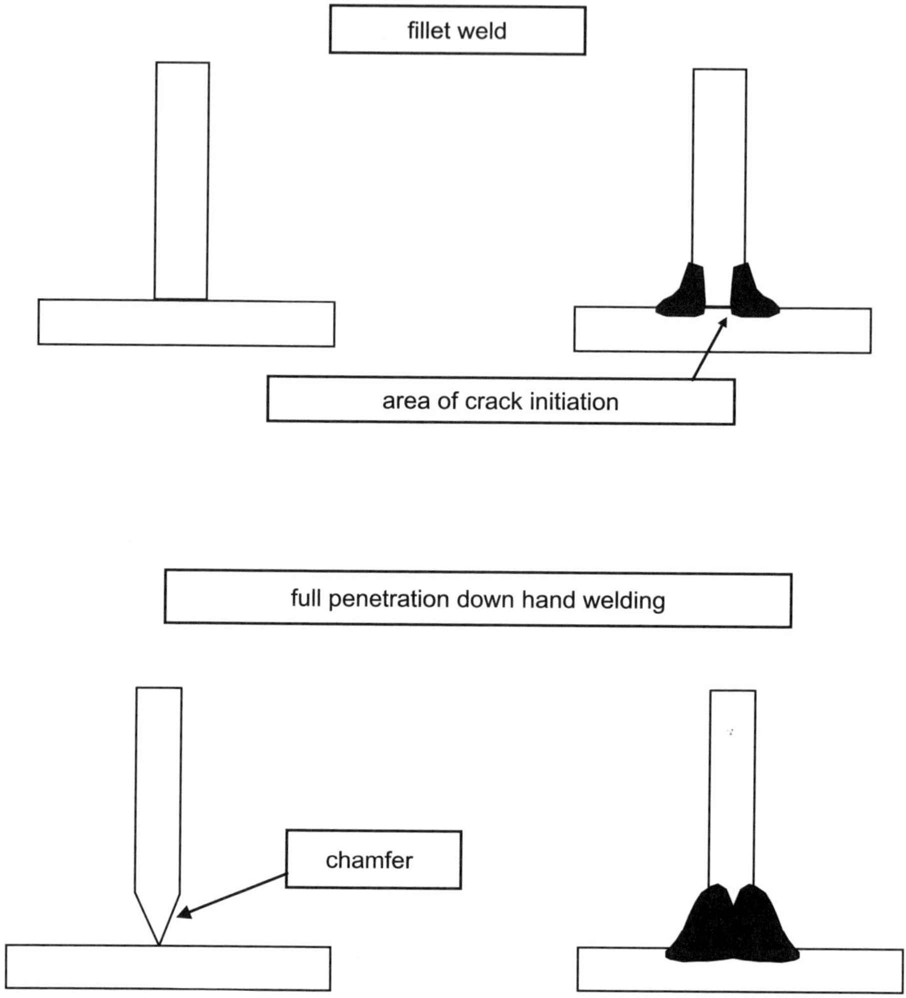

Fig. 15.4. Welds—fillet and full penetration

chamfered before welding. This increases the cost. All critical full-penetration, down hand welding usually is X-rayed.

The cylinder block usually is made up of castings. The cylinders are individual. There is one cylinder head for each cylinder. Each cylinder contains its own cylinder liner. The cylinder also contains one half of the passage for the tie rod. At the end of the engine, a special casting with one half a tie rod passage is bolted on.

Some modern slow-speed engines use a round combination water jacket distance piece as part of the cylinder. Longer cylinder head studs hold the cylinder head, cylinder liner, and water jacket distance piece to the lower part of the cylinder.

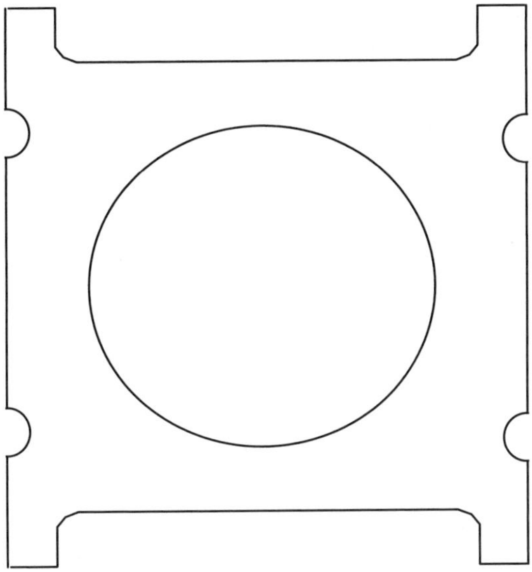

Fig. 15.5. Top view of cylinder

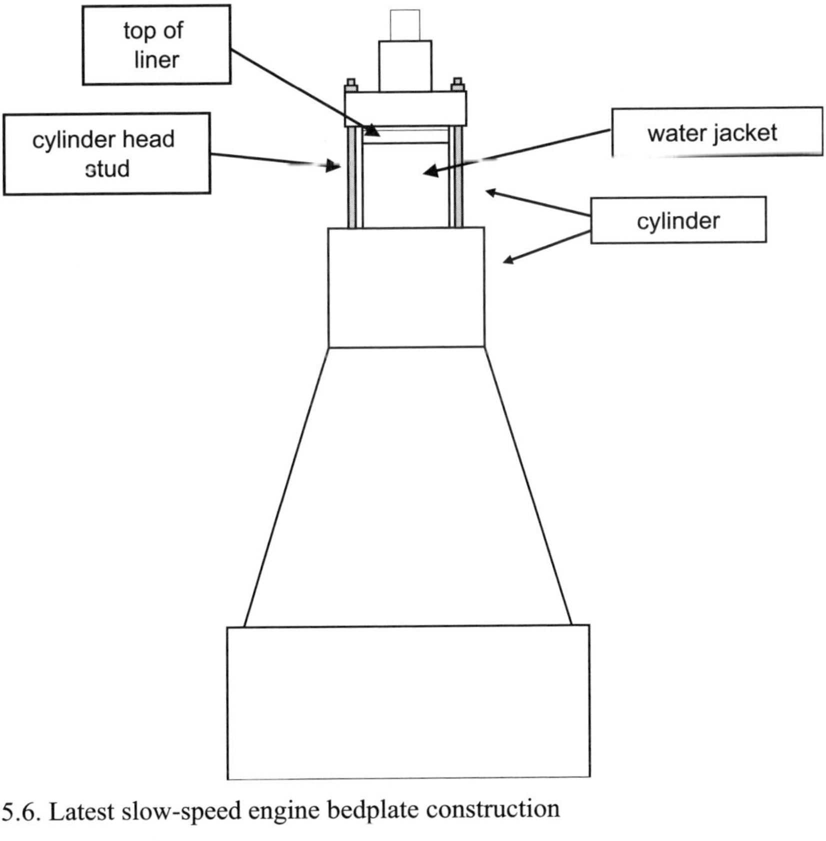

Fig. 15.6. Latest slow-speed engine bedplate construction

Another type of engine construction is the block construction. This is limited to small, high-speed engines because the engine frame is a cast, one-piece block. This block contains the cooling water passages and the cylindrical passages for the cylinder liners and the bearing supports and hollows for the crank throws and connecting rods.

This is a very cost-effective method to produce an engine frame, but it is limited to smaller engines because the shrinkage of the solidifying molten metal causes voids in a large casting.

An oil pan is bolted to the bottom of the block to contain the oil sump. A one-piece head is bolted to the block for small engines. Larger, high-speed engines have a block and separate heads for each cylinder.

One problem is the under slung crankshaft. If the bearing cap bolts become loose, the crankshaft has no support at this cap. With the bedplate type construction the cap is on the top. The main engine forces will not be pushing on the cap with bedplate construction, but will be pushing on the cap with block construction.

The trend in medium-speed engine construction is to use a weldment. This is essentially a block welded up of steel plates. Full penetration down hand welding is used. Welding sequences and procedures are very important to minimize warping. The weldment is an-

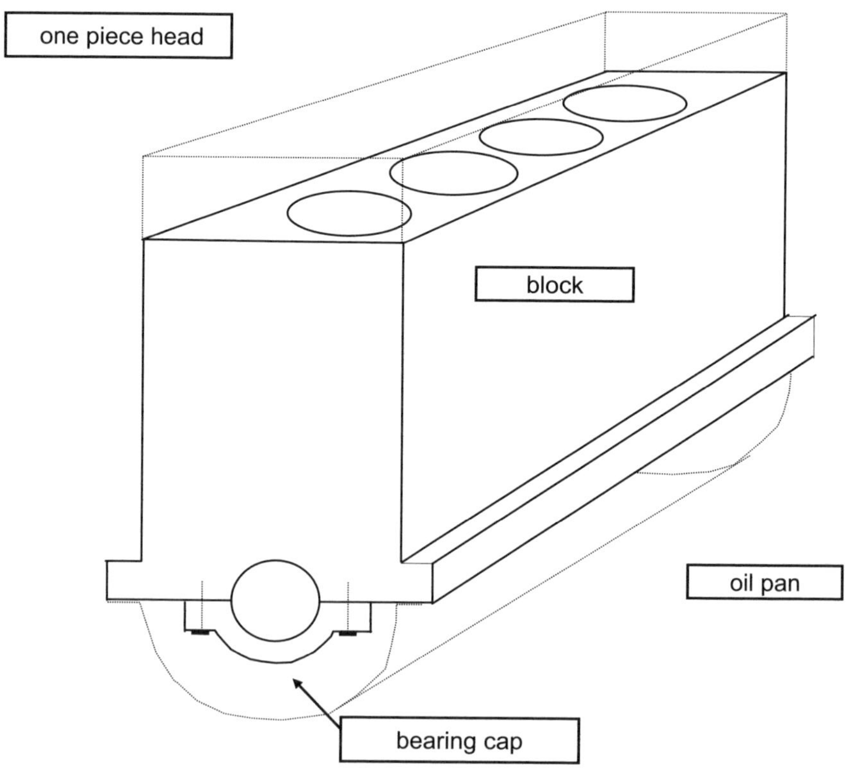

Fig. 15.7. Block construction

nealed and the final critical dimensional surfaces are machined. This manufacturing technique reduces the cost as compared with bedplate construction. Most medium-speed engines are too large to have a one piece cast block.

This weldment is madeup of plates that form a box around each cylinder bay. This is necessary to ensure a very strong box girder arrangement that contains a bearing support for each main bearing of the crankshaft. These plates may be up to two inches thick. Being essentially a block type construction, we are still left with the under slung crankshaft. The horizontal and vertical bearing housing bolts help to ensure proper support for the crankshaft. Since the block weldment is so large, it has room for this number of bearing support bolts.

A water jacket is first put in the cylinder hole shown fig. 15.8. Inside the water jacket, the wet liner is inserted. The cylinder head is placed on top of the cylinder liner. It is held in place by cylinder studs screwed into the weldment. Thus, any parts subject to corrosion or wear can be replaced.

Even though the engine frame is made as strong and stiff as possible, the ship's structure must also help keep the crankshaft bearings aligned. Steam turbines are very short when compared to a main propulsion diesel engine. The turbines are connected to the reduction

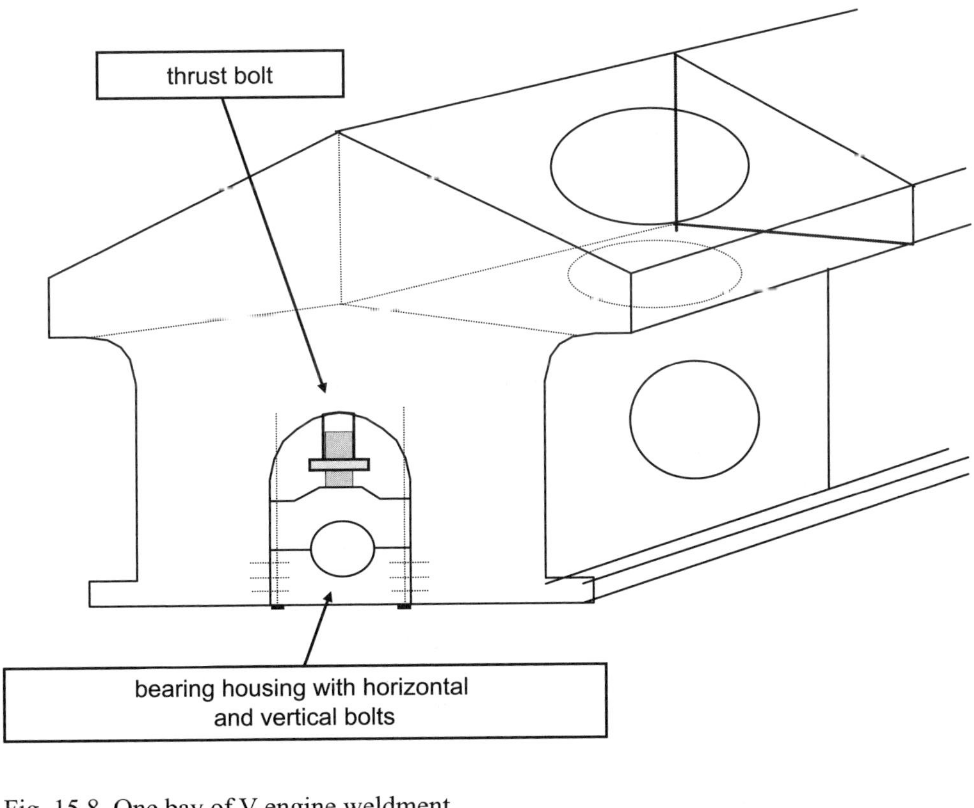

Fig. 15.8. One bay of V-engine weldment

gears with quill shafts. Thus, only a short distance must be kept in close alignment before it makes a flexible connection. The crankshaft of a main propulsion diesel engine has no such connection throughout its length. Some slow-speed engines can have a bedplate more than eighty feet long. The crankshaft bearings must be kept in very close alignment. The structure under the bedplate or weldment must be stiffer and stronger as compared to a steam turbine ship.

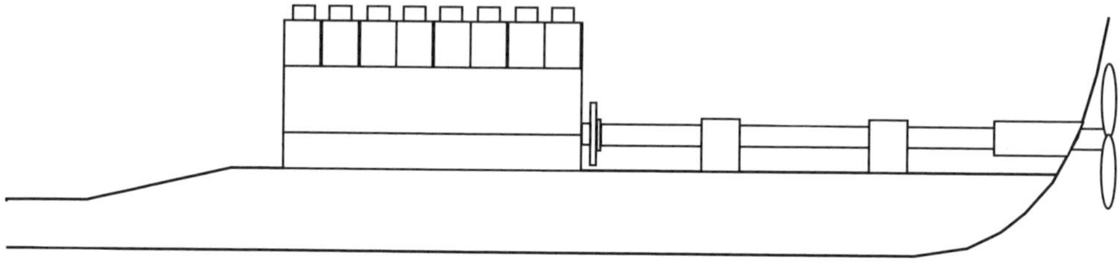

Fig. 15.9. Ship structure under engine room

This is accomplished by having the double bottoms under the engine room deeper and made up of thicker plating. The transition to the regular double bottoms under the rest of the ship must be tapered so as not to induce a stress riser in the ship.

The engine crankshaft bearings can come out of alignment during groundings and unusual loading of cargo. This is why crankshaft deflection readings should be taken periodically, and especially after a grounding.

This part of the ships structure underneath the main propulsion diesel engine is welded, using full-penetration down hand welding. Most engine rooms are located in the aft part of the ship in modern construction. This means that the longitudinal members disappear as the ship narrows toward the stern. This is taken care of by jogging the longitudinal ensure is enough stiffening structure is near the narrowing stern section.

Main propulsion diesel engines have the lubricating oil sump located in the double bottom below the engine. A void area known as a cofferdam is located around this sump. The foundation bolts holding the bedplate or weldment to the ship are located through the plating over the cofferdam. The foundation bolts can be checked on both sides by going into the cofferdam through a manhole. Any leaks around the foundation bolts will end up in the cofferdam instead of the lubricating oil sump.

The cofferdam also contains level indicating instruments, a sounding tube, and a vent. At the bottom of the sounding tube, a tee is installed. One end of the tee connects to the sump, and the other end has a replaceable plug to absorb the impact of the sounding weight.

The foundation bolts are longer than needed to just go through the bedplate or engine frame and cofferdam tank top. Most hold down bolts are as long as possible to keep the stress in the bolts from becoming too great. The shorter the bolt, the quicker the stress rises with elongation as the bolt is torqued. Torquing usually means using a torque wrench. On larger engines a hydraulic bolt stretcher will be used.

Chocks are used between the bedplate and the tank top. They usually are designed to be about 1.5 inches thick. This allows for alignment of the engine and the reduction gears

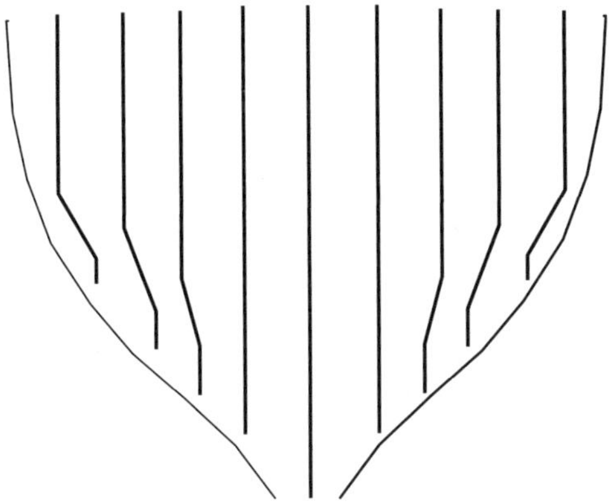

Fig. 15.10. Jogged longitudinal members near stern

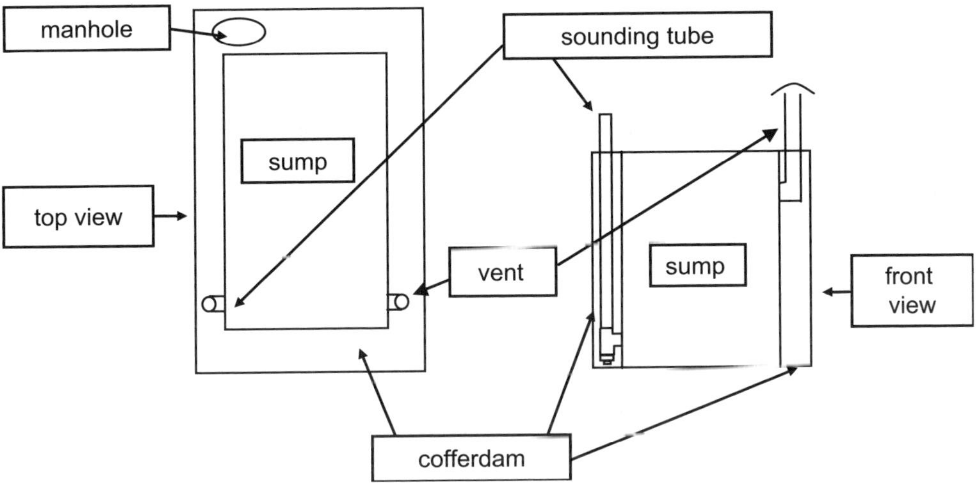

Fig. 15.11. Lubricating oil sump and cofferdam

or propeller shaft. After the ship is built and the engine is being installed, the engine is lined up with the reduction gears or propeller shaft with jack screws. Then the chocks are machined to the exact thickness to align the engine. Today, instead of using machined steel chocks, epoxy resin may be used. The engine is lined up and foam rubber dams are inserted around the foundation bolts and epoxy resin is poured in.

During engine operation, the foundation bolts should be periodically checked. This can be done by retorquing. The bolt also can be tapped with a hammer while holding your thumb nail on the opposite side of the bolt. You will notice any looseness by the bolt banging on your thumb nail as you tap the bolt.

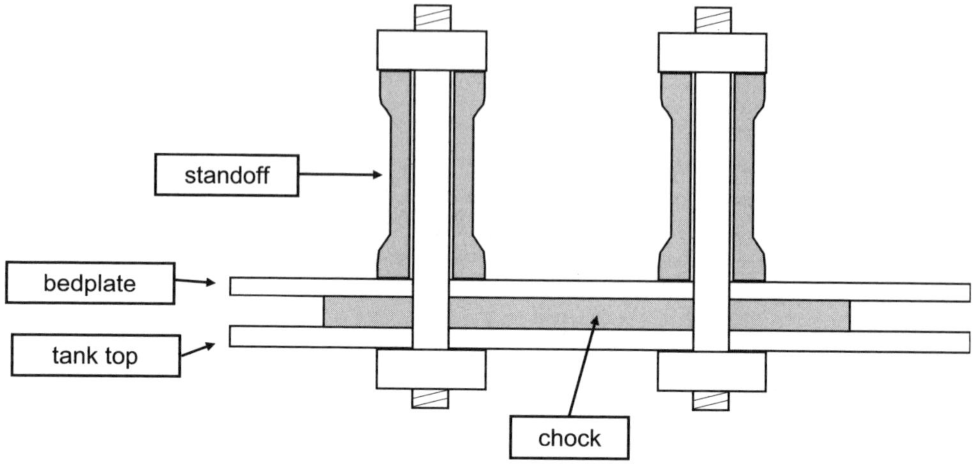

Fig. 15.12. Foundation bolts

Watch for fretted chocks. This condition occurs when the chock, tank top, or bedplate wears due to looseness and vibration. If a fretted chock goes unnoticed and the foundation bolt is retorqued, the crankshaft bearings can come out of alignment. Fretted chocks usually have a rusty powder around the area of fretting. If the area is moist and oily, the rusty powder may appear as a reddish grease. To correct for this, the chock is removed and the area cleaned. The engine is realigned and the chock usually is replaced with an epoxy resin chock.

Study Questions

1. In diagram A, which component supports the crankshaft?
   A. 1
   B. 2
   C. 3
   D. 4

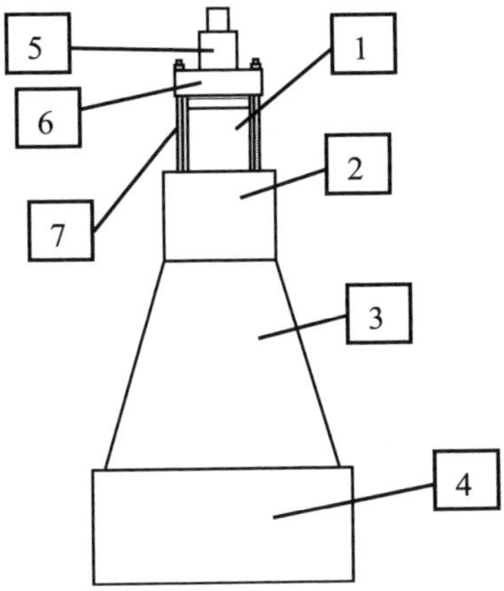

Diagram A

2. In diagram A, which component houses the crosshead?
   A. 1
   B. 2
   C. 3
   D. 4
3. In diagram A, which component represents the exhaust valve?
   A. 1
   B. 5
   C. 6
   D. 7
4. In diagram A, component 7 is a:
   A. head stud
   B. tie rod
   C. stay rod
   D. tie stud
5. In diagram A, which component represents the cylinder?
   A. 1
   B. 2
   C. 1 and 2
   D. 5 and 6

6. The engine represented in diagram A is what type of engine?
   A. inline medium speed
   B. slow speed
   C. high speed
   D. opposed piston
7. High-speed engines use what type of basic construction?
   A. bedplate
   B. weldment
   C. columns
   D. block
8. Chocks can be made of what material?
   A. steel
   B. epoxy
   C. ceramic
   D. A or B
9. Where is the lube oil sump located in a main propulsion engine?
   A. bedplate
   B. double bottom
   C. weldment
   D. oil pan
   C. 1 and 2
   D. 5 and 6

Chapter 16

Running Gear (Crank Shafts, Connecting Rods, and Bearings)

Many parts of a diesel engine are subject to cyclic loading, because the maximum firing pressure only happens during a short period of crankshaft rotation. This makes fatigue an important design factor in many components of the diesel engine. The fatigue limit is found by testing materials under cyclic loads of different values. A S-N diagram is then constructed. The S stands for stress and the N stands for the number of cycles to failure.

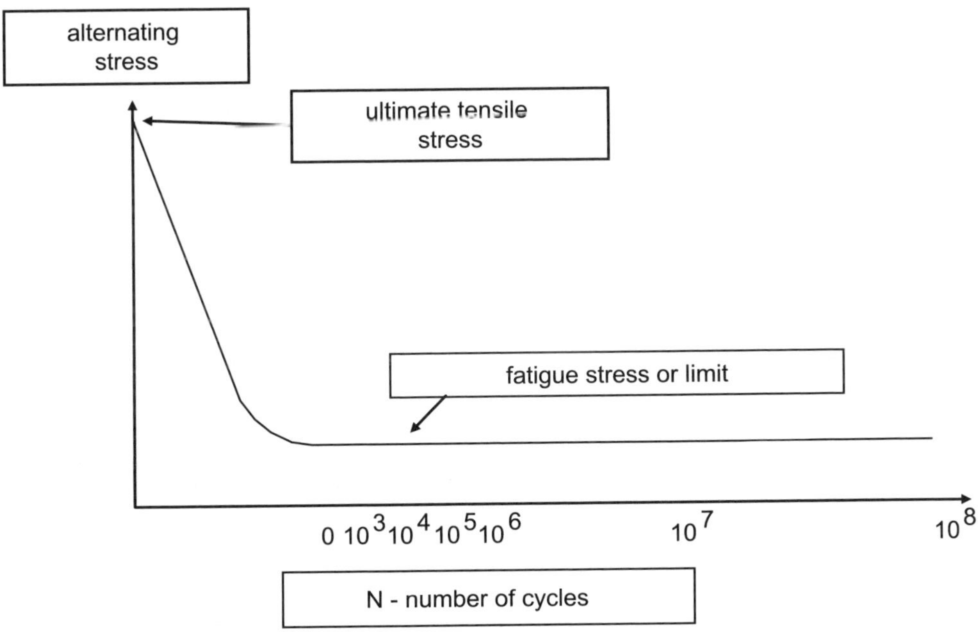

Fig. 16.1. S-N or stress to number of cycles to failure

If the maximum load is supplied to the test specimen in one cycle, we have the ultimate tensile stress. As we decrease the load and apply it at more cycles, we eventually reach the fatigue limit or strength. As you can see, the fatigue limit is much less than the ultimate tensile stress.

Any stress riser or stress concentration factor will make the situation worse. This can include just a ding on the surface of a tie rod or connecting rod. This is especially true for shafting, since the maximum stress is on the surface of the shaft. Any component in bending has the same problem of the maximum stress on the surface where a ding could occur.

All the journals in the same crankshaft are the same diameter. The journal near the output shaft has the highest stress but all the components of the crankshaft are the same dimensions for balancing and ease of manufacturing.

Fillets are needed between the journals and webs to avoid the sharp corner that would form a stress riser. The diagram above shows a regular fillet, but some crankshafts use a reverse fillet. The reverse fillet is machined into the web instead of being incorporated into the journal. This saves space and shortens the crankshaft. The rule of thumb is that the radius of the fillet equals 5 percent the diameter of the journal. The outlet and inlet of the oil passages also must be radiused for the same reason. The rule of thumb is the radius is equal to the diameter of the oil passage. The surface of the journal is the most highly stressed area of the round shafting. Since the oil passages pierce the surface of the journal, this is an especially critical area. Cracks usually start from an oil passage on the surface of the journal of an overloaded crankshaft for medium and high-speed engines. Slow-speed engines do not have oil holes drilled in the crankshaft.

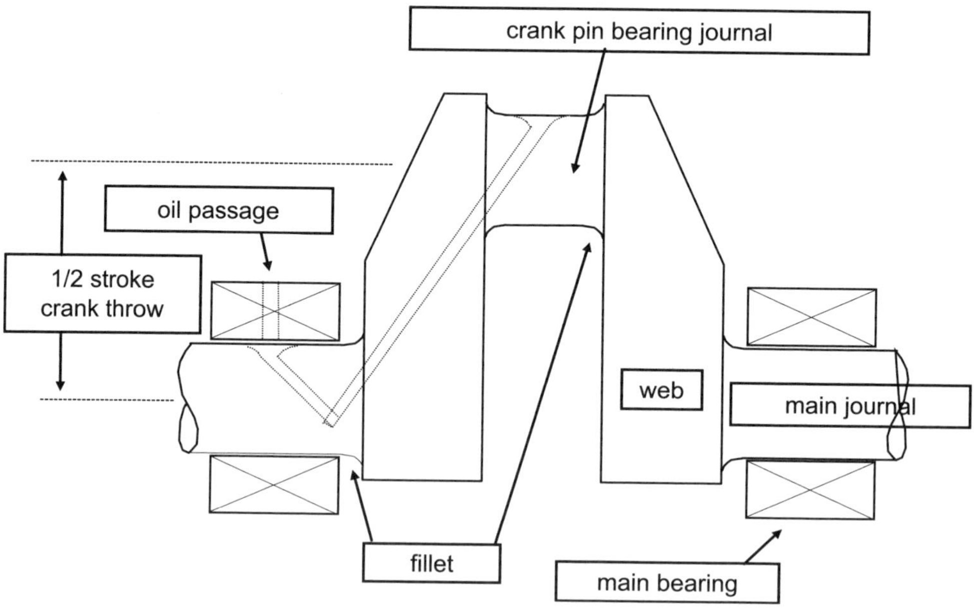

Fig. 16.2. Crankshaft

All crankshafts for high- and medium-speed engines are usually built in one piece. Smaller engines use a casting. Smaller engines can use cast iron or cast steel. Larger, higher output engines are made of an alloy steel forging. This includes medium-speed propulsion engines.

Slow-speed engines are so large that the crankshaft is builtup of components. The earliest slow-speed engines used a fully builtup crankshaft. Only two components were necessary, webs and journals. The latest version of this type of crankshaft uses a shrink fit to keep the components together.

A shrink fit is used instead of keyways or teeth, because these discontinuities would induce stress risers in the joint between the web and the journal.

The problem with the construction in fig 16.3 is that it made for a heavy crankshaft. Extra material is needed in the web for the shrink fit. This led to the semi builtup crankshaft.

The reason this type of construction can be used was the development of the technology to forge such a large piece that included both the crank pin and the webs.

Again the shrink fit is used. The difference between the hole and the journal is about 1/570 to 1/660 the diameter of the journal insert diameter.

The latest evolution is the crankarm.

This single cast or forged piece can makeup any crankshaft. The reason it has only recently come into production is because of advances in welding technology. These arms have to be welded in the middle of the journal from the inside out.

As shown in fig. 16.5, this makes the lightest, slow-speed crankshaft because no shrink fit material is needed.

Some crankshafts need counterweights to balance the engine. The crankshaft must be balanced both statically and dynamically.

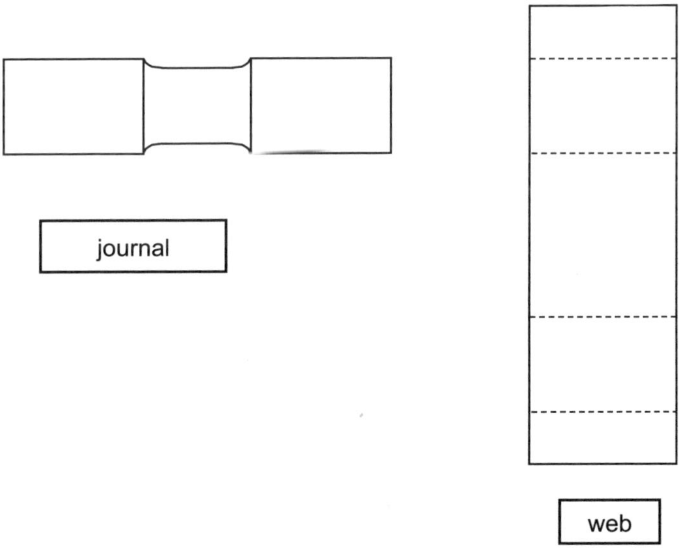

Fig. 16.3. Fully builtup crankshaft components

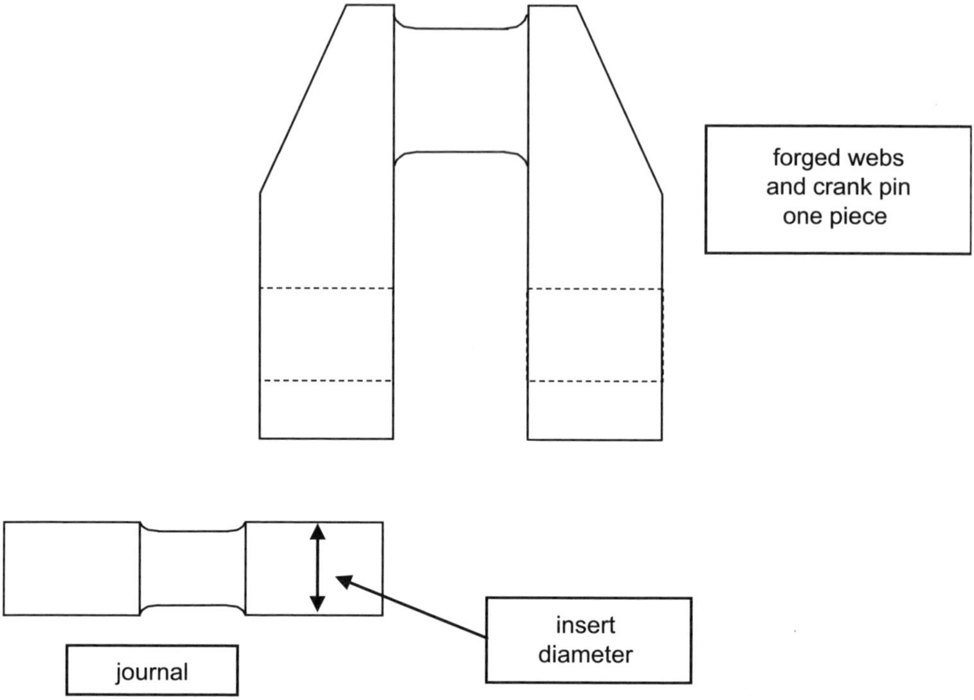

Fig. 16.4. Semi-builtup crankshaft components

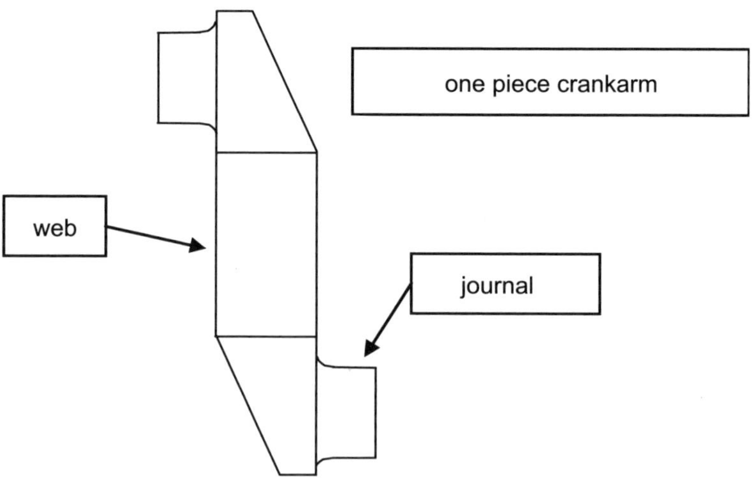

Fig. 16.5. Crankarm crankshaft component

To dynamically balance an object in rotation, at least two weights must be added in two different planes. Counterweights on the crankshaft can be made integral with the webs or attached to the webs. If the counterweights are bolted to the webs, take care not to over-stress the crankshaft in the bolt holes. These bolt holes introduce a discontinuity in the crankshaft and, a stress riser. Colt-Pielstick uses the configuration in fig. 16.7 to get around this problem.

This configuration (fig. 16.7) also makes for easy removal of the counterweight and crankshaft.

The journal surfaces of the crankshaft are very important for long life. Large, slow-speed engines have the surfaces ground. Smaller engines have the journals ground and surface hardened. This makes for the best high-speed wearing surface that is hard on the surface, but tough underneath. Otherwise, a completely hardened journal will be strong but brittle. The surface hardened zone must completely include the fillet or it must stop before the fillet. If it ends in the fillet, we have a geometrical and material discontinuity in the same spot, giving us a stress riser.

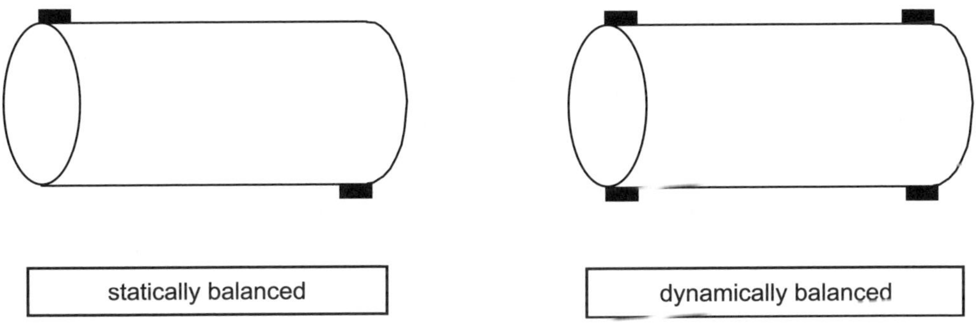

Fig. 16.6. Balancing

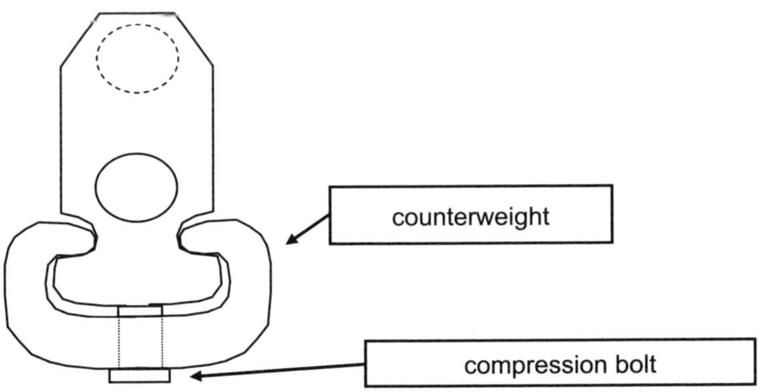

Fig. 16.7. Counterweight attachment with no stress riser

The crankshaft design is based on the torque transmitted to the output shaft and just as importantly, fatigue. Most of the load on the crankshaft is cyclic in nature. These loads include:

- Working cycle. The variable gas pressure in the combustion chamber.
- Torsional vibration. Induced by the combustion pulses turning the crankshaft.
- Axial vibration. A smaller vibration set up as the crankshaft vibration, like an accordion.
- Misalignment. A rotating shaft that is bowed induces tension and compression, alternating stresses on the top and bottom of the shaft as it rotates.

It also is important to know the natural frequency of the crankshaft and make sure it is not in the normal operating range.

All these cyclic loads are primarily a design problem, except misalignment, which can be caused by bearing wear, ship grounding, and uneven cargo loading. The misalignment can be found by using crankshaft deflection readings taken by the operator.

In fig. 16.8, the low or worn bearing would cause the webs to open. If the bearing was high, the webs would close. The dial gauge usually is set at zero at TDC. Readings are taken at port, starboard, and two readings at BDC on either side of the connecting rod. The dial gauge is placed at the same punch marks in the web every time. The readings should not vary more than + or - .025 mm (.001 in.), except near the flywheel, where they may be a little greater.

The crankshaft on some medium-speed engines is stiff enough the low bearing do not affect the crankshaft deflection readings. This is called a bridged bearing. Some manufacturers require a hydraulic jack to load the crank throw at TDC to see if the bearing is bridged.

Bearing wear can be detected using a bridge gauge on large engines. On small engines the bearing shell can be taken out and measured with a bearing micrometer. This entails taking the bearing apart. Some manufacturers have designed a fixture for a depth microm-

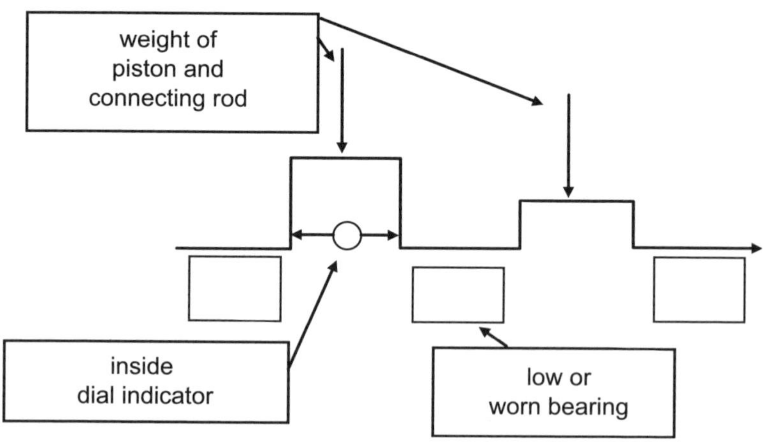

Fig. 16.8. Crankshaft deflection readings

eter that fits between the bearing cap and the web. This fixture rests on the engine frame. Thus, bearing wear down can be measured without disassembly.

Thermal growth of the crankshaft also must be considered. The crankshaft is very long and it will grow considerably as the engine warms to operating temperature. In slow-speed engines, the crankshaft is fixed at the flywheel end by the thrust bearing. The rest of the crankshaft is allowed to grow and contract with temperature. The main bearing supports and the webs have enough lateral clearance between them to take care of this. In high-speed and medium-speed engines, the thrust bearing is in the reduction gears. The last bearing shell near the flywheel is U-shaped and acts like a locating bearing for the crankshaft.

The only part of the crankshaft that can be reconditioned is the journal. The journals are usually made oversize so that they can be reground. When doing this, it must be remembered to blend in the fillets. Sometimes the journal may be chrome plated or metal sprayed for reconditioning to original diameter. The problem is fatigue failure may occur at the coating material interface.

The bearings used in diesel engines are journal bearings. The larger bearings use shells for easy maintenance. All the bearings except the piston pin bearing and the crosshead bearing rotate. The bearings work on the principal of a hydrodynamic oil wedge, keeping the rotating journal separated from the stationary shell.

The film pressure inside the hydrodynamic oil wedge does not depend on the system oil pressure. The film pressure depends upon the interaction between the journal and the shell. This pressure can go as high as 1,000 psi or more.

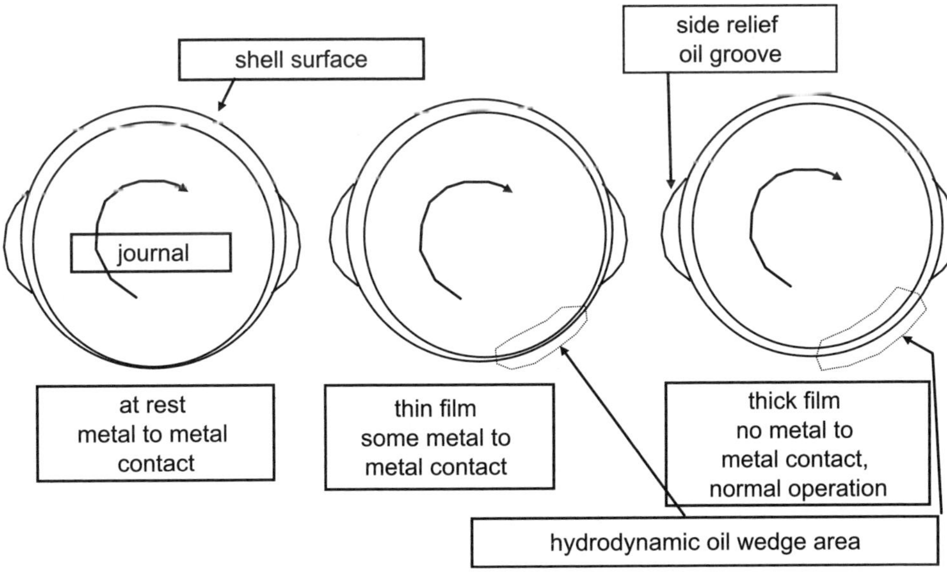

Fig. 16.9. Journal bearing operation

The film pressure depends upon:

1. Oil viscosity. The higher the viscosity, the higher the film pressure.
2. Temperature of the oil. The higher the temperature, the lower the oil viscosity that lowers film pressure.
3. Journal speed. The journal acts as an oil pump to create the hydrodynamic oil wedge.
4. Bearing clearance. The bearing clearance determines the confinement of the oil. The smaller the clearance, the higher the film pressure. The larger the clearance, the lower the film pressure. However, higher oil flow with this larger clearance, removes fictional heat generated by the shearing of the oil. The oil molecules on the journal are rotating with it while the oil molecules on the shell are stationary. The molecules in between act as molecular balls generating heat as they keep the journal separated from the shell. A rule of thumb is .001 of an inch clearance for each inch of diameter of the shaft. The bearing clearances can be measured with feeler gauges on large engines. Plastigauge is used on smaller engines. Plastigauge is similar to lead wire except the wire is made of plastic. This tends to lessen problems of bearing metal indentations. The plastic wire is placed between the bearing journal and the top shell. The top cap is then torqued down. The plastic wire deforms, taking the thickness of the bearing clearance. A comparator gauge is supplied to measure the width of the flattened plastic to thousands of an inch thickness.
5. Bearing load. The higher the bearing load, the smaller the clearance in the oil wedge area and the higher the film pressure. This makes the bearing is self regulating with load. However, if the journal touches the bearing shell during normal operation, the bearing will fail. This is managed by having enough bearing area so the journal does not come in contact with the bearing shell.

Normally, some oil grooves in the bearing distribute oil so that the hydrodynamic oil wedge can be formed. These grooves must be well rounded so that the edge of the groove does not wipe oil away from the journal. Side relief oil grooves are usually the best, because they are not located in the hydrodynamic oil wedge area. The pressure in the groove is essentially zero compared with the film pressure. The pressure at the edge of the bearing is also essentially zero. Thus, we have the pressure distribution as seen in fig. 16.10.

The bearing without the oil groove can support more load. The bearing half where the oil wedge is formed usually does not have an oil groove. It can, but the bearing width must be increased to support the load. Some manufacturers have the bearing shell half where the oil wedge is not located, grooved to distribute oil, and the bearing half where the oil wedge is located, not grooved. If these halves are inadvertently switched, the bearing will fail.

Four-stroke engines have another problem. The journal is moving around in the bearing clearance as well as rotating. This is due to the load reversal during intake and exhaust.

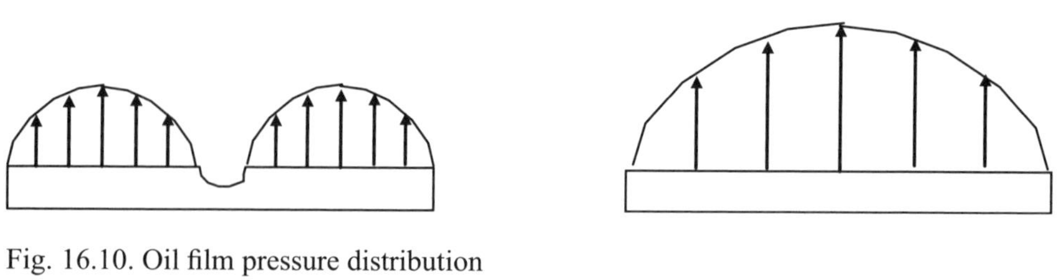

Fig. 16.10. Oil film pressure distribution

This phenomena can cause the relative motion of the journal in the oil wedge area to reverse. This can momentarily destroy the oil wedge and eventually destroy the bearing. This can be fixed by adding counterweights to the crankshaft to diminish this motion.

White metal or babbitt is mostly tin and lead with some antimony. The proportions change, depending on the desired hardness. Antimony is the constituent that hardens the most. This alloy is placed on a steel back. This is known as a bimetal bearing.

A trimetal bearing has a copper-lead alloy bonded to the steel back. Then the babbitt is bonded to the copper-lead alloy.

Some manufacturers use aluminum bearings. They can have either an aluminum coating on a steel back or a solid aluminum-tin shell.

Modern high output engines use precision-bearing shells. This means that the babbitt is very thin—.013 to .025 of an inch on small engines. On older engines, that babbitt could be 1/8 inch thick. The thick babbitt allows the bearing to fit with scrapers and shims. The thick babbitt could be squeezed out the sides of the bearing in high-output engines. Today, the precision bearing with thin babbitt is used. Because the babbitt is thin, it acts like a coating and is not easily squeezed out. The babbitt lets the journal slide over the shell during startup and shutdown, when metal-to-metal contact occurs. The babbitt also absorbs particles in the oil that would damage the journal. With the thin babbitt on precision bearings, the size of the particle that can be absorbed is smaller. Thus, purifiers and filters are very important. Because of the thin babbit, no precision bearing can be custom fitted to the journal. When the precision bearing shell is worn, it must be replaced.

With assembling bearings, you must adhere to the following:

1. The bearing shells are slightly oversized. When the cap is torqued down, we have nip or crush to ensure the shell is in intimate contact with the housing. The shell is not a strength member; the bearing housing is the strength member. The shell is just a replaceable liner. Its shape is dictated by the bearing housing. Do not file the ends if the shell looks a little large.
2. The shells have some type of antirotation device. This will be either a pin or a tang. The tang fits in a slot in the half of the housing it is located in. The other half of the housing has no slot. A pair of tangs prevents the shell from rotating. However, if the

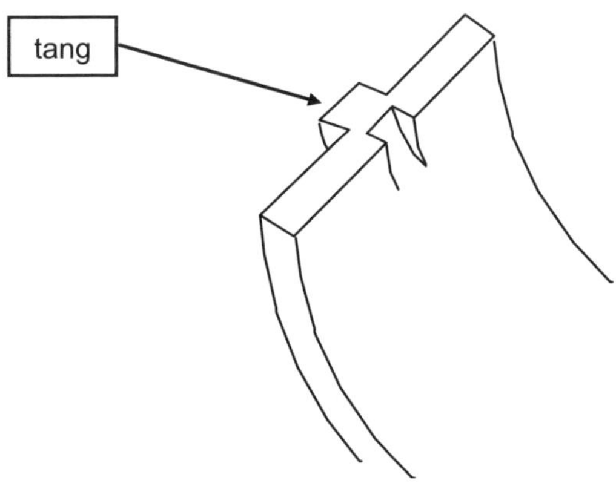

Fig. 16.11. Bearing shell tang

shells are not matched with the correct half of the housing, the bearing shell edge will be pushed into the journal.

3. The bore of the housing and the back of the bearing shell must be perfectly clean. The smallest amount of dirt can lead to a high spot and bearing failure.
4. The piston must be in correct alignment with the crankshaft. Otherwise, the crank pin bearing will be edge loaded and fail.
5. Make sure the bearing shell halves are not swapped because of the antirotation device and the oil grooves.

Piston pin bearings and the crosshead bearing rock instead of rotate, which allows very little chance of a hydrodynamic oil wedge to form. These bearings depend upon squeeze pressure to separate the journal from the shell. Four-stroke engines and high- and medium-speed, two-stroke engines have a load reversal near TDC. This helps draw in oil to the alternating clearance as the load reverses during the working cycle.

The slow-speed two-stroke engine has very little load reversal. Elevated oil pressure is directly led to the crosshead bearing by an articulated oil line.

The basic idea is to have oil under the journal before the fuel combusts. The journal is separated from the lower shell or insert. The combustion pressure pushes the two pieces toward each other, but just before they touch the combustion pressure is expanded out and gone. With a load reversal in the four-stroke engine, the oil is replaced. With the slow-speed engine, the system oil pressure forces oil between the journal and shell.

The slow-speed engine has special oil grooves in the lower shell to help the oil get in between the crosshead journal and the lower shell.

The main reason for the crosshead engine is to eliminate the piston pin bearing and bring it out of the piston to increase its size. The forces are much larger in a slow-speed engine because of the slow speed. A much larger bearing area is needed at the top of the

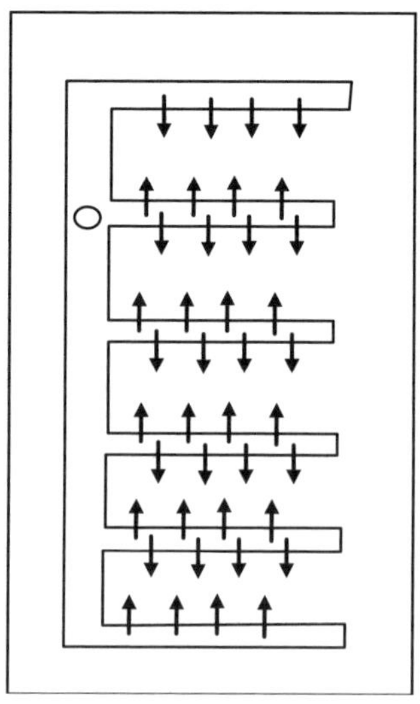

Fig. 16.12. Top view of crosshead bearing shell with oil grooves

connecting rod for the squeeze pressure to keep the journal and bearing surfaces apart during operation of the engine.

Because of the articulated oil line, the flow of oil to the crank pin bearing is down the connecting rod for the crosshead engine. The trunk piston engine has the oil flow going up the connecting rod to lubricant the piston pin bearing.

Connecting Rod

The connecting rod is essentially a pinned-end column that primarily is in compression, which could buckle at the slender part of the connecting rod. The large compression load occurs during the power stroke. With a four-stroke engine, the connecting rod is in tension between the intake and exhaust strokes. Because the load is cyclic, the connecting rod must be designed for fatigue. It also must be designed for buckling or elastic instability.

For pinned ends, the Euler buckling formula is:

P is the maximum load

$$P = \frac{\pi^2}{l^2} EI$$

E is the Modulus of Elasticity

I is the moment of inertia of the cross section of the column

l is the length of the column

The generic formula for the moment of inertia is $I = k^2 A$.

The significance of these two formulas is that the larger the moment of inertia of the connecting rod's cross section, the larger the maximum load can be. The larger the radius of gyration (k) for a given cross sectional area, the larger the moment of inertia.

The geometry with the highest moment of inertia for a given area is the I-beam. That is why the I-beam was invented. All medium- and high-speed engines use connecting rods made of forged steel with an I-beam cross section. Slow-speed engines can use a circular cross section because of the low speed and the large size of the connecting rod. Also, it is easy to manufacture a connecting rod with a circular cross section. It is much more expensive to forge a connecting rod with an I-beam cross section, but with medium- and high-speed engines this has to be done. Besides the compression load, the connecting rod is whipping back and forth. This tends to bow the connecting rod momentarily and promote buckling. With the medium- and high-speed engines, the only cross section and manufacturing technique that can be used is forging with an I-beam cross section.

The outside surfaces of the connecting rods usually are polished on high-speed engines. This is to help eliminate stress risers on the surface of the connecting rod. It is very important not to nick or ding these surfaces and induced a stress riser.

The marine type connecting rod was used when the big-end-bearing shells had a thick babbitt. The babbitt could be scraped and fitted as the bearing shell wore. The split on the lower end of the connecting rod at the bearing shells could be shimmed to adjust the bearing clearance. Since the bearing was scraped or just wore down, the piston would drop, decreasing the compression ratio. Shims could be added under the palm to raise the piston

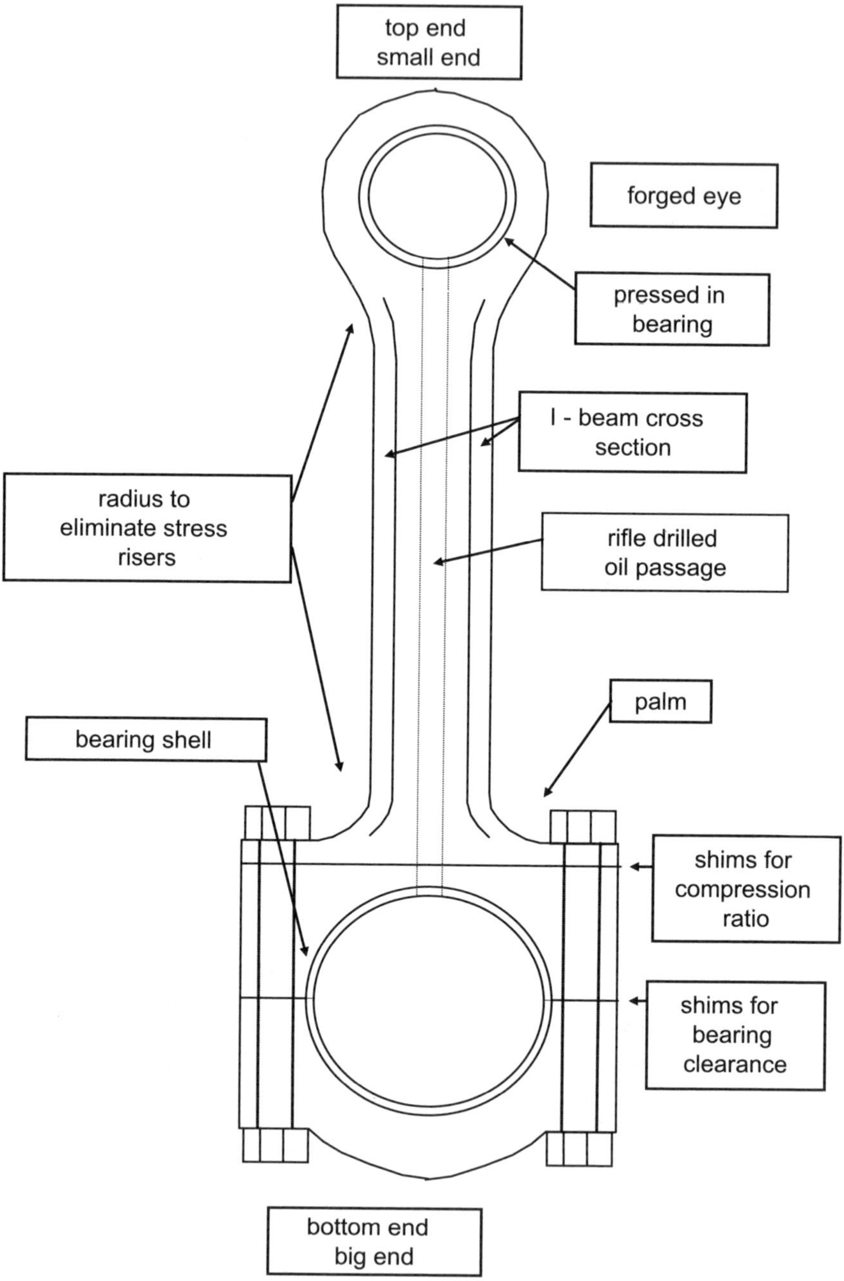

Fig. 16.13. Marine-type connecting rod

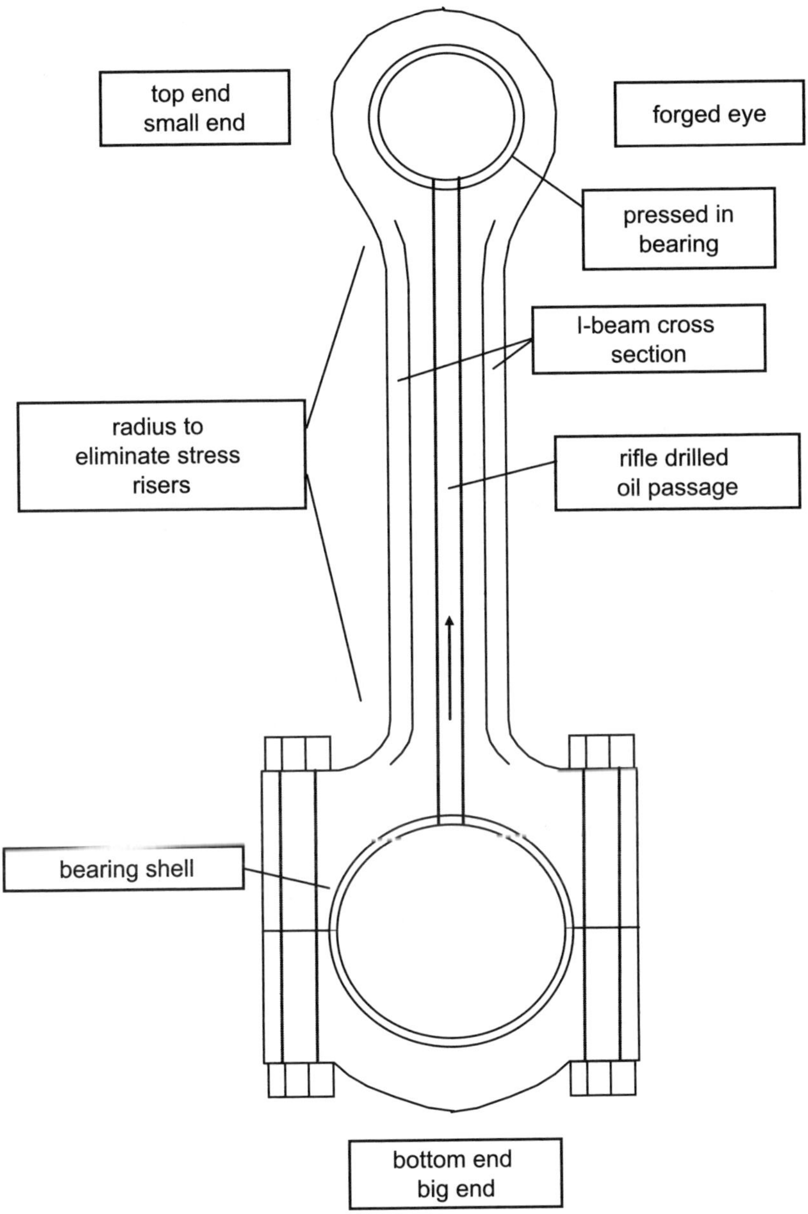

Fig. 16.14. Automotive-type connecting rod

back to its original specifications. Today, however, precision-bearing shells with thin babbitt are used. If the babbitt wears, then the shell has to be replaced. The extra joint is not needed in the connecting rod. Many modern engines use an automotive type connecting rod for this reason. Some slow-speed engines still use the marine type connecting rod with precision bearings to keep the weight of the individual parts of the connecting rod down to a reasonable level.

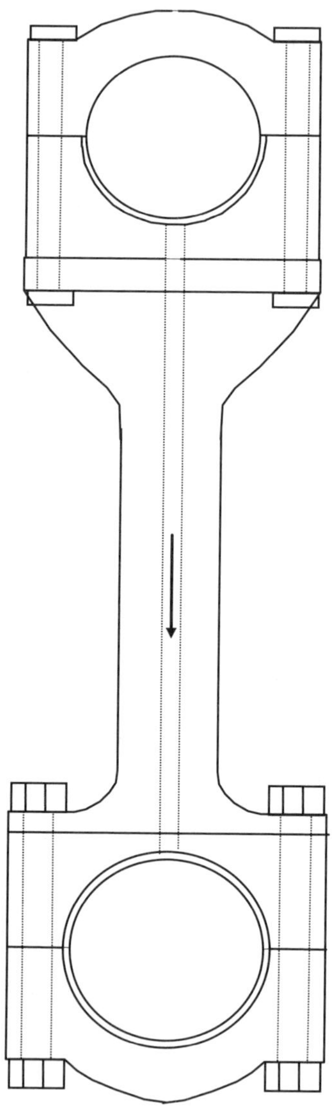

Fig. 16.15. Crosshead connecting rod

The slow-speed engine connecting rod is not only different because of the circular cross section, but because the upper end of the rod ends in the crosshead bearing. This means that the upper end is forked and has a joint and bearing cap. This must be done to allow disassembly of the connecting rod from the crosshead journal. It is forked also to allow for the piston rod to be connected to the crosshead bearing.

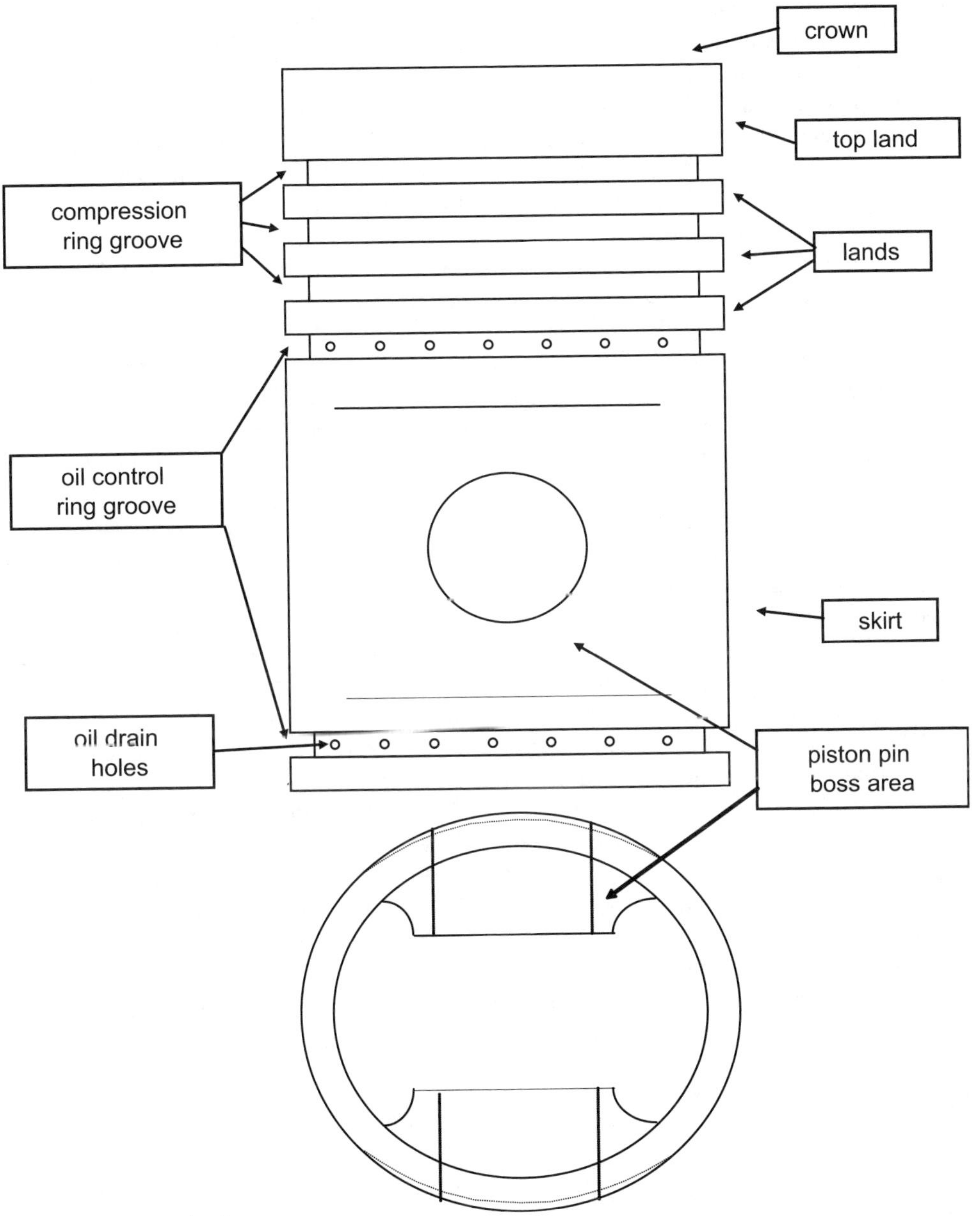

Fig. 16.16. Trunk piston

There usually is only a lower bearing shell for the crosshead bearing. The cap has very little load on it because this is a two-stroke cycle engine. An oil hole through the middle of the connecting rod supplies oil to the crankpin bearing, since oil is supplied to the crosshead bearing by the articulated oil line. The oil flow is down the connecting rod for a crosshead engine and up the connecting rod from the crankshaft in a trunk piston engine.

The trunk piston is tapered or step cut with the smaller diameter at the top. When the piston comes up to operating temperature, it is the correct shape.

Some pistons are cam ground in the way of the piston pin bosses. These bosses are extra metal that transfer the gas load from the piston wall to the piston pin and onto the connecting rod. Since this makes the piston wall thicker, it expands more than the thinner piston wall as the engine comes to operating temperature. The cam grinding allows expansion of the piston pin bosses without reducing the clearance between the piston and the liner. Instead of cam grinding, the piston may be elliptical in shape, with the minor diameter containing the piston pin bosses.

The number of compression rings varies inversely as the speed of the engine. The slower the engine, the more rings are needed to contain the pressure.

Oil control rings are needed so that only the correct amount of oil reaches the upper part of the liner. Some oil must reach the top compression ring for lubrication. Some of this oil will be burned off due to the high combustion temperatures. Too much oil reaching the top ring would be a waste of money. Too much oil in this hot area could also cause carbon deposits and a stuck ring. Wear and the amount of oil used are a compromise. More oil would cause less wear until deposits form.

The crosshead piston, also known as a barrel piston, has only compression rings. No oil control rings are used because the cylinder lubrication system provides once through oil to the piston through the lubricating oil quills. These quills contain a check valve and an accumulator piston with spring. Oil is injected into the liner during periods of low pressure. This can happen when the first compression ring goes by or when the piston is at TDC. The oil is supplied by small, positive displacement piston pumps. If the pumps are not timed to the piston movement, then the check valve and accumulator piston are needed.

The crosshead piston also is tapered with the smaller diameter at the top to obtain the correct shape as the engine comes up to operating temperature.

These pistons require internal cooling. Loop-scavenged engines use cooling water supplied through two telescopic tubes. These are shown dotted in fig. 16.17. The loop-scavenged piston also needs a so-called skirt to keep the scavenge ports closed while the piston is at TDC. This also is shown dotted fig. 16.17. The uniflow-scavenged engine does not require either one of these features. Crankcase oil is used to cool the piston and it is supplied through the piston rod.

Water is needed to cool the loop-scavenged engine because exhaust gas passes over the piston during blowdown. In the uniflow-scavenged engine, only cool intake air blows across the piston, thus, requiring less cooling. This allows for oil cooling with its associated lower heat transfer coefficient.

Notice that the piston is relatively short compared to a trunk piston. This is because the trunk piston has to absorb the side thrust caused by the angles that the connecting rod makes during engine operation. The crosshead piston rod is bolted to the crosshead bearing and the connecting rod is attached to this journal. The crosshead journal contains a linear

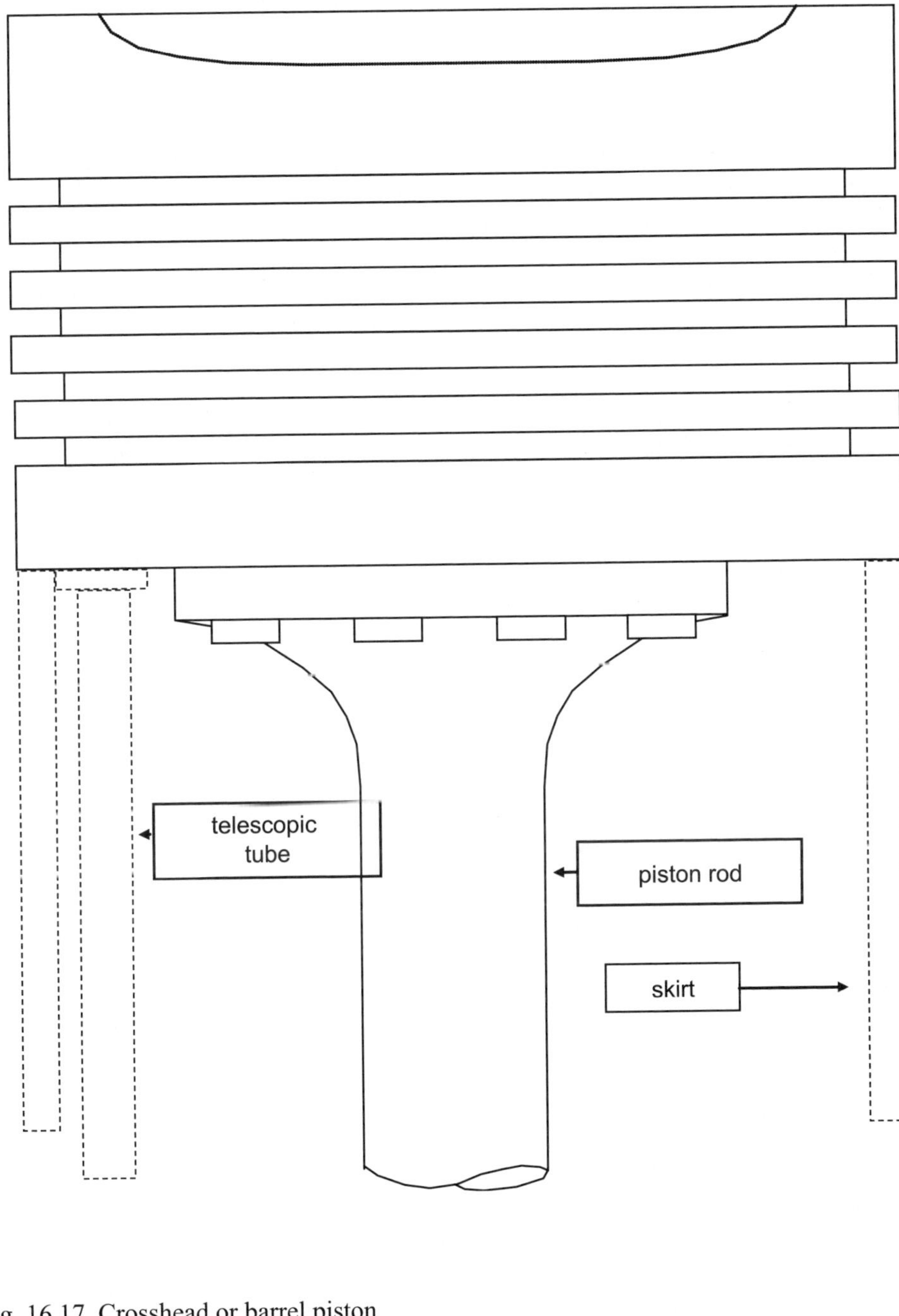

Fig. 16.17. Crosshead or barrel piston

slipper bearing at each end that absorbs the side thrust generated by the connecting rod. The crosshead piston does not absorb any side thrust.

Piston pins and the crosshead journal are made of forged steel. They are surface hardened, ground, and micro-finished. If you drag your fingernail across the surface, you should feel no resistance to movement.

The three kinds of piston pins are:

1. Full floating. This is the most common type of piston pin. It is free to rotate in the piston and in the top of the connecting rod.
2. Semi-floating. With this type of piston pin, the connecting rod is bolted to the pin with more bearing area located in the piston.
3. Fixed. This type of piston pin has the pin bolted to the piston with more bearing area located in the connecting rod.
4. Semi-floating and fixed piston pins are seen in some applications. They are designer schemes to increase the bearing area in either the piston or the connecting rod, but not both.

Rotating Piston for Trunk Piston Engines

The piston pin and piston pin bosses present a problem in uneven thermal expansion and stress distribution. Sulzer (Wartsila) has eliminated the piston pin in some of its medium-speed, trunk-piston engines. It has been replaced by a ball and socket joint. A pawl arrangement is built into the socket below its center. An associated ring gear is built into

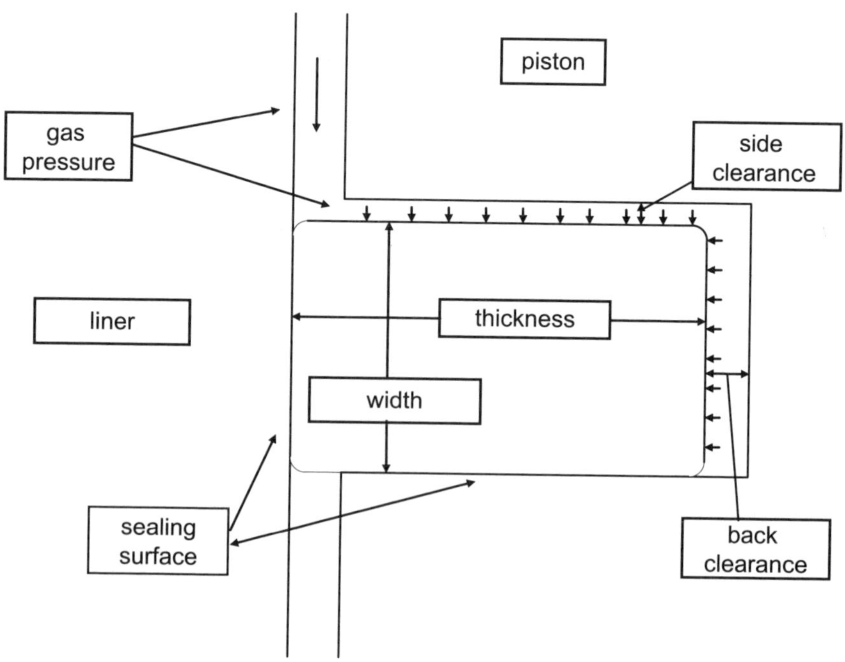

Fig. 16.18. Compression ring sealing

the piston opposite the pawl arrangement. The connecting rod rocks back and forth, very slowly rotating the piston.

This has the following advantages:

- A newly oiled portion of the piston skirt is always available to absorb side thrust.
- Improved oiling and even wear of the piston rings.
- Because the upper part of the connecting rod is spherical and the piston is symmetrical, the material of both components are more evenly loaded.
- Less thermal distortion because of the symmetrical piston.

Compression Rings

The compression ring depends upon gas pressure to push down on and behind the ring. This produces a force on the sealing surfaces proportional to the gas being sealed against. To ensure the compression ring starts in this position, the ring is made larger than the bore. This pretension in the compression ring is called dimetral tension. Each compression ring also has an end gap to allow for the ring to be installed on the piston. This is done with a ring expander. This tool expands the ring by putting an even bending moment on the ring to prevent ring breakage. After the rings are installed, a ring compressor is needed to slide the rings and the piston into the bore.

The top compression ring is subject to the highest pressures and temperatures. It also has the highest pressure drop across the ring. The top ring must be kept below 200 °C to ensure the lubrication film does not break down. This is why the top land is so big. In the

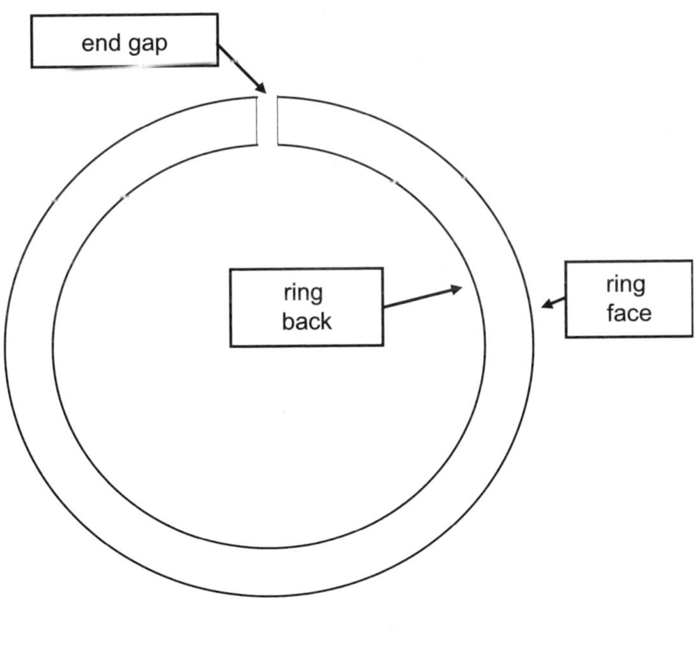

Fig. 16.19. Top view of compression ring

time it takes the hot gas to get between the water cooled liner and the oil cooled piston, the temperature will have dropped enough so as not to destroy the lubricating oil film.

It is very important to keep the clearances between the piston ring groove and the compression ring within proper tolerances. The side clearance must be big enough to allow the gases to push the ring against the sealing surfaces. If the side clearance is too large, excessive ring hammer and excessive oil pumping action occurs. This oil pumping action happens because the compression ring goes up and down in the groove during engine operation. This movement actually pumps oil up into the combustion chamber. If it has too little side clearance, the gas pressure cannot get behind the ring. This can cause an elastic instability collapse of the compression ring, because gas pressure is on the face of the ring instead of behind it, pushing the compression ring against the liner.

The back clearance must be large enough to allow the gases to get behind the ring and promote sealing. Too large a back clearance allows too much gas flow. This does not allow the gases to cool as they travel behind the ring. Too large a back clearance also allows too much oil pumping. Both of these conditions allows oxidation of the oil and carbon packing behind the ring, eventually leading to a stuck ring.

The end gap is important because if it is too small the ring will jam in the liner as the engine comes up to temperature. If the gap is too large, excessive gas leaks past the compression ring. The end gap is checked in a new liner with a feeler gauge. It may also be checked in an unworn portion of the cylinder being worked on. It is important to check this dimension in an unworn portion of the liner. If it is checked in a worn portion of the liner and a ring is chosen with this end gap, the end gap will be too small as the compression ring travels toward BDC and the unworn part of the liner.

The lower side of the compression ring and the mating side of the piston ring groove must be perfectly smooth. If they are not, you have an unsupported area of cast iron ring with a load on top. This leads to a broken compression ring, and the ring will leak.

Compression and oil control rings are made of flake graphite cast iron. This material absorbs oil and with the cast iron liner and makes a very good wearing surface that helps

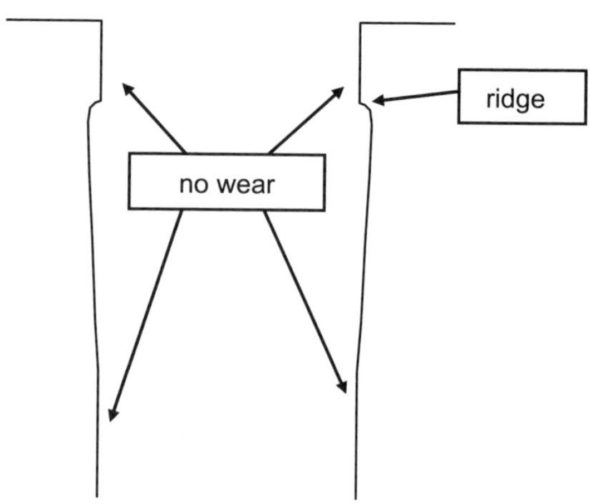

Fig. 16.20. Liner wear pattern

with its own lubrication. To strengthen the cast iron, alloying elements such as manganese, chromium, and molybdenum are added. Since the top ring is subject to the highest temperatures and pressures, it may be chrome plated.

Piston rings wear because of the relative motion between the ring, the piston and the liner. The wear pattern of the liner also contributes to the relative motion between the side of the ring and the side of the ring groove. The ring diameter is increasing and decreasing constantly with piston movement.

Since the top ring does not travel all the way to the top of the liner, a ridge forms. This ridge must be ground to blend in with the unworn part of the liner when new compression rings are installed. Otherwise, the top compression ring could break.

The piston also moves from side to side during side thrust reversal in four-stroke engines. This causes relative movement of the side of the compression ring and the side of the piston ring groove.

All this wear is the reason that the piston rings are the most short-lived components of a diesel engine. During the maintenance of a cylinder, the liner is measured with an inside micrometer and the piston is inspected. The side clearance of the compression rings can be measured with a feeler gauge. The thickness of the ring can be measured and the back clearance calculated. However, since so much work is consumed pulling the piston, the usual practice is just to replace all the rings. The time interval for pulling a piston is based on the expected piston ring wear limit.

After new piston rings are installed, a run-in period at reduced load must be observed. Some parts of the compression ring face will not be in intimate contact with the cylinder wall. If the engine is brought to full power too quickly, hot gases blow through these small spaces, destroying the lubricating oil film and causing microseizures between the compression ring and the liner. If the load is brought up in a prescribed manner, the ring seats and microseizures are eliminated. Some compression rings have a bronze insert or coating to help with a quick run-in time.

The piston ring grooves in aluminum pistons may wear abnormally because of the relative softness of aluminum as compared with cast iron. A special Ni-resist cast iron insert containing the piston ring grooves may be used. This special alloy has a high coefficient

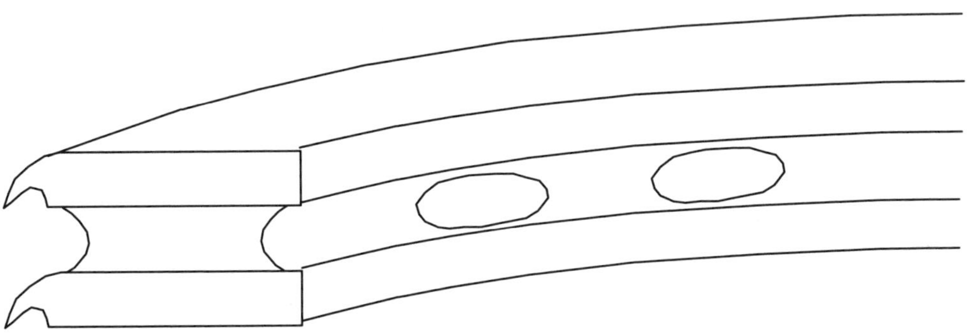

Fig. 16.21. Ventilated oil control ring

of expansion compared with regular cast iron. This ensures that the cast iron insert stay bonded to the aluminum piston.

Oil control rings are only used on trunk piston engines. These rings ensure that only the correct amount of oil reaches the upper part of the liner for the lubrication of the compression rings and liner. Without these oil control rings, excessive amounts of oil will be consumed, due to the oil pumping characteristics of the compression rings and the vast amount of oil being thrown up from the crankpin bearing. Most oil control rings are ventilated scraper rings as shown in fig. 16.21.

Oil is scraped from the liner and travels through the ports in the ring and then through the holes in the back of the oil control ring groove in the piston. These holes are shown in the trunk piston diagram. The holes in the back of the oil control ring groove lead to the inside of the piston and back to the crankcase.

Very little gas pressure is available to push the oil control rings against the liner. With the above ventilated rings, the pressure is equalized and some type of spring is used to push the ring against the liner. The spring runs completely around the back of the oil control ring. It may be a helical or a flat spring. The double scraper face reduces the contact area with the liner to make full use of this spring force.

Study Questions

1. The main journals of a running crankshaft have the highest stress located:
   A. the surface
   B. the center
   C. halfway between the center and the surface
   D. three-quarters the way between the center and the surface
2. Where do cracks usually start in an overloaded four-stroke cycle engine crankshaft?
   A. at the centerline of a journal near an oil passage
   B. at the surface of a journal near an oil passage
   C. inside the web
   D. on the surface of a web
3. Which type of crankshaft is only possible with advanced welding technology?
   A. forged one piece
   B. fully builtup
   C. semi builtup
   D. crank arms
4. Crankshaft deflection readings do not show problems associated with which of the following?
   A. a grounding
   B. unusual cargo loading
   C. a bridged bearing
   D. a worn bearing
5. What is the approximate pressure in the hydrodynamic oil wedge in psi?
   A. 10
   B. 100
   C. 1,000
   D. 10,000
6. What is an antirotation device for a bearing shell?
   A. fiction fit
   B. tang
   C. tab
   D. stake
7. Which engine has oil flowing down through the connecting rod?
   A. high speed
   B. medium speed
   C. slow speed
   D. trunk piston
8. 8. Connecting rods normally fail in what mode?
   A. tension
   B. compression
   C. elasticity instability
   D. plastic instability

9. What cross sectional area has the most area away from the neutral axis?
   A. circular
   B. rectangular
   C. elliptical
   D. I-beam
10. In diagram A, a split bearing shell is represented by:
   A. 1
   B. 6
   C. 1 and 6
   D. none of the above
11. Shims are added to change the bearing length at what point in diagram A?
   A. 4
   B. 5
   C. 4 and 5
   D. none of the above

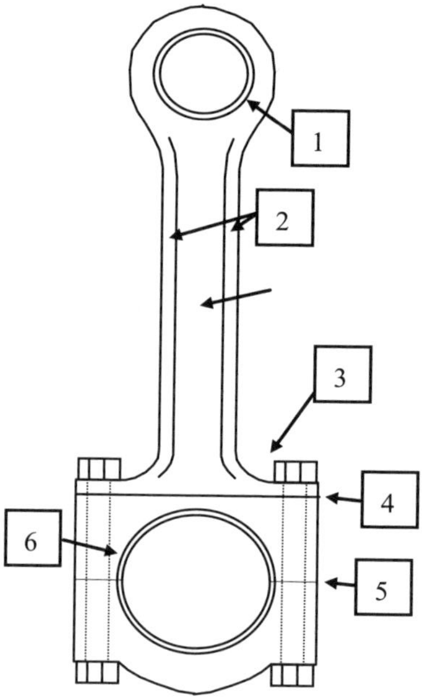

Diagram A

12. The object in diagram A represents?
   A. an automotive type connecting rod
   B. a Marine type connecting rod
   C. a crosshead type connecting rod
   D. a double-link connecting rod

13. Shims are added to change the compression ratio at what point in diagram A?
    A. 4
    B. 5
    C. 4 and 5
    D. none of the above

14. Oil is supplied to the top of the object in diagram A:
    A. from the lower end
    B. from an articulated oil line
    C. from oil thrown up from the crankshaft
    D. by the piston lubricators

15. In diagram B, the gas pressure force is transmitted to the piston pin by:
    A. 3
    B. 4
    C. 5
    D. 6

16. In diagram B, which diameter is manufactured the smallest?
    A. two
    B. three
    C. four
    D. they are all the same

17. In diagram B, which number helps prevent the compression ring from overheating?
    A. 1
    B. 2
    C. 3
    D. 4

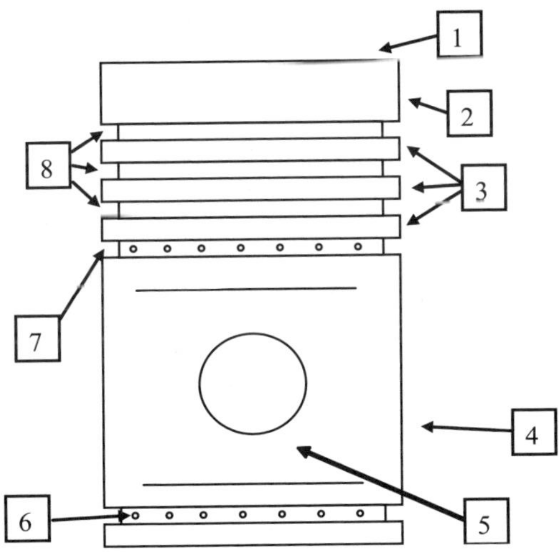

Diagram B

18. How many oil control rings are used in diagram B?
    A. one
    B. two
    C. three
    D. five

19. In diagram B, which number is the crown?
    A. 1
    B. 2
    C. 3
    D. 4

20. Which engine will have the least oil control rings?
    A. high speed
    B. medium-speed four-stroke
    C. slow-speed four-stroke
    D. slow speed

21. Which piston is hardest to manufacture?
    A. trunk
    B. barrel
    C. rotating
    D. oscillating

22. What pushes a compression ring against the liner?
    A. gas pressure
    B. oil pressure
    C. spring force
    D. centrifugal force

23. What part or parts of a compression ring are sealing surfaces?
    A. face
    B. back
    C. side
    D. A and C

24. What pushes an oil control ring against the liner?
    A. gas pressure
    B. oil pressure
    C. spring force
    D. centrifugal force

25. What piston ring clearance is critical in preventing a compression ring failure in elastic instability?
    A. end
    B. back
    C. side
    D. A and B

26. What determines the time interval between piston pulls?
    A. liner life
    B. piston ring life
    C. piston skirt life
    D. piston crown life

27. An S-N diagram has to do with:
   A. fatigue
   B. acid neutralization
   C. ultimate tensile stress
   D. neutralization value

28. How does the diameter of the journals vary along the length of the crankshaft?
   A. largest at the flywheel end
   B. smallest at the flywheel and
   C. largest at the camshaft drive
   D. same throughout

29. Most slow-speed engines in operation today have what type of crankshaft?
   A. forged one piece
   B. fully builtup
   C. semi builtup
   D. crank arms

30. The object shown is:
   A. statically balanced
   B. symmetrically balanced
   C. dynamically balanced
   D. unbalanced

31. What generates the hydrodynamic oil wedge?
   A. interaction between journal and bearing shell
   B. crosshead oil pump
   C. lube oil service pump
   D. oil grooves

32. What type of bearing shells are used in modern medium-speed engines?
   A. thick walled
   B. thin walled
   C. precision
   D. exact

33. Which bearing uses squeeze pressure in its operation?
   A. crosshead
   B. crankshaft journal
   C. crankpin journal
   D. thrust bearing

34. Connecting rods act as a column with what type of ends?
   A. pinned
   B. fixed
   C. jointed
   D. articulated

35. Which engine will have the most compression rings?
    A. High speed
    B. Medium-speed four-stroke
    C. Medium-speed two-stroke
    D. Slow speed

36. The crosshead bearing in a slow-speed engine could be categorized as what type of piston pin in a trunk piston engine?
    A. full floating
    B. semi floating
    C. fixed
    D. stationery

37. Which piston may be water cooled?
    A. trunk
    B. barrel
    C. rotating
    D. oscillating

38. What tool is needed to insert the piston and rings into the cylinder bore?
    A. ring expander
    B. ring compressor
    C. ring extender
    D. ring extracter

39. What may be applied to a compression ring face to promote rapid run in?
    A. bronze coating
    B. chrome plating
    C. Ni-resist coating
    D. hard coating

40. Piston rings are made of what material?
    A. spring steel
    B. stellite
    C. cast iron
    D. bronze

41. Piston pins are made of what material?
    A. forged steel
    B. cast iron
    C. aluminum
    D. cast steel

Chapter 17

Crankcase Explosions

To have a crankcase explosion, the atmosphere in the crankcase must be in the flammability range for the air-fuel ratio. The temperature must be above the autoignition point. During normal conditions, the air-fuel ratio is much too lean, meaning not enough hydrocarbons are mixed in with the air. The flash point of lubricating oil is usually more than 400 °F. The temperatures in the crankcase usually are well below this point. Thus, crankcase explosions are extremely rare in a well-maintained engine.

A hot spot is needed to vaporize the lubricating oil to get the air-fuel ratio in the flammability range and provide the high temperature needed for ignition. Hot spots can be caused by the following:

1. Hot bearing. This could be any bearing in the crankcase from the main journal bearing to a roller chain sprocket bearing.
2. Hot roller chain guide. MAN-B&W uses a large roller chain to drive the camshaft. This cause of hot spots is mainly related to slow-speed engines. The hot bearing can be for crosshead and trunk piston engines. Crosshead engines have the piston rod seal. Any problems with the piston will not reach the crankcase. However, problems with the piston will reach the crankcase in a trunk piston engine.
3. Incipient trunk piston seizure. Just as the piston is about to seize, friction will generate plenty of heat for a hot spot. Once the piston has seized, everything cools off.
4. Perforated trunk piston crown. A small hole in a trunk piston crown allows a combustion flame to enter the crankcase. If the hole is too big, there is no compression to ignite the fuel and no flame.

Blowby in trunk piston engines do not cause a crankcase explosion. Sparks have an extremely difficult time passing by all the compression and oil control rings to get to the crankcase, even if they are in terrible condition.

The hot spot vaporizes the lubricating oil and sets up the correct air-fuel ratio for an explosion in a local area. If the hot spot sets off the oil and air mixture, a flame front starts and combusts any oil vapor that is near an oxygen molecule. This happens even if the air-fuel ratio is not within the range of flammability. This initial primary explosion causes more lubricating oil to vaporize and raises the pressure inside the crankcase to about 5 psi. This is not a very large explosion. However, this small pressure rise acts on a large area, creating a large force that could crack open the crankcase. If this happened, it would allow oxygen to rush in, and react with the newly generated oil vapor from the primary explosion. This creates a very large secondary explosion that could cause considerable damage. It is the secondary explosion that kills people in the engine room.

Designed into the engine are a number of safety features that minimizes the dangers of a crankcase explosion:

1. Explosion relief doors. These are essentially relief valves set to pop at about .1 bar or 1.4 psi. This relieves the pressure from the primary explosion and prevents the more destructive secondary explosion. The explosion relief doors have a built-in flame arrestor and flame shield. This protects personnel. Each crank bay should have an explosion relief door.
2. Small openings between crank bays. If the opening over the main crankshaft journals is small, then the flame front will not pass through it, igniting the vapor in the rest of the crankcase.
3. Crankcase ventilation. A small crankcase exhaust blower is used to keep the crankcase at a negative pressure and the air-fuel ratio above the range of flammability.
4. Catwalk location. Catwalks are located to deflect flames away from operating personnel as seen below (fig. 17.1).
5. Mist detectors. A fine white mist of oil is created by the hot spot. This is due to the explosive boiling of the lubricating oil by the heat of the hot spot. The same phenomena occurs when oil is dropped on a hot exhaust manifold. The so-called white smoke is an oil mist. An oil mist detector, like the Kiddy system, is used to detect flammable oil mist inside the engine crankcase. Air from various bays of the crankcase passes between a light and a photocell. The mist reduces the light getting to the phototube and sets off an alarm.

If the mist detector alarm goes off or a primary explosion occurs, the engine must be shut down immediately and the engine room must be evacuated. It is recommended that twenty minutes pass before personnel enter the engine room to inspect the crankcase. If the

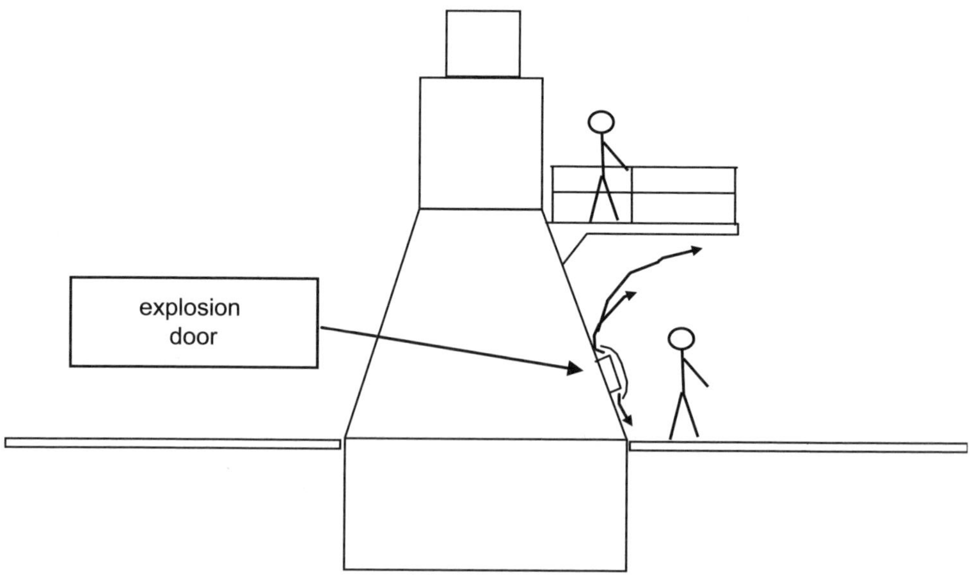

Fig. 17.1. Crankcase explosion gas deflection

crankcase is opened immediately, the air currents could produce an area near the hot spot in the range of flammability and cause an explosion. When inspecting the crankcase, look for the hot spot. This may be difficult to find because it could be any bearing or rubbing surface in the crankcase.

In one case a primary explosion occurred in slow-speed engine. The crankcase was inspected after twenty minutes and nothing found. The engine was started, and after coming up to load, another primary explosion occurred. The cause eventually was found to be a ship's hot thrust bearing located in the after part of the crankcase.

Study Questions

1. Explosion relief doors prevent:
   A. primary explosions
   B. hot spots
   C. secondary explosions
   D. oil mist
2. What causes the most damage in a crankcase explosion?
   A. the primary explosion
   B. the secondary explosion
   C. the flame front
   D. the hot spot
3. An explosion relief door opens at what pressure differential?
   A. .1 psi
   B. .1 bar
   C. 1.4 bar
   D. 14 psi
4. What is a cause of a hot spot in a slow-speed engine?
   A. hot bearing
   B. blowby
   C. incipient piston seizure
   D. perforated piston
5. What gives an indication that a crankcase explosion is about to occur?
   A. crankcase lube oil temperature
   B. cylinder lube oil temperature
   C. flame detector
   D. mist detector alarm

About the Author

Professor Daniel P. Charnews is a Full Professor of Marine Engineering and has taught at the United States Merchant Marine Academy for the past twenty-nine years. He is a past recipient of the Alice B. McNulty Award for Excellence in Distinguished Teaching. Previously, he taught Marine Engineering at the Massachusetts Maritime Academy. He holds a Bachelor of Marine Engineering with Honors from the New York Maritime College and an Ocean Engineers Degree from the Massachusetts Institute of Technology and the Woods Hole Oceanographic Institution. He coauthored several papers on diesel engine combustion with Dr. Robert Latorre of the University of New Orleans for the *Journal of Ship Research* and presented a paper on Diesel Engine Room Simulation in Nantes, France. He is a member of the American Society of Engineering Educators and the Society of Naval Architects and Marine Engineers. He holds a First Assistant Engineers License, Unlimited Horsepower Steam and Diesel from the United States Coast Guard and is a registered Professional Engineer in the State of New York. He is the father of two grown children and lives on Long Island, New York, with his wife Barbara.

Bibliography

Christensen, Stanley G., *Lamb's Questions and Answers on the Marine Diesel Engine, 8th Edition,* London: Edward Arnold, 1990.

Cowley, J., *The Running and Maintenance of Marine Machinery,* London: The Institute of Marine Engineers,1992.

Harrington, R. L., *Marine Engineering,* Jersey City: The Society of Naval Architects and Marine Engineers,1992.

Henshall, S.H., *Medium and High Speed Diesel Engines for Marine Use,* London: The Institute of Marine Engineers,1978.

Hunt, Everett C., *Modern Marine Engineer's Manual, Volume II,* Centreville, Maryland: Cornell Maritime Press Inc., 1991.

Knak, Christen, *Diesel Motor Ships Engines and Machinery,* Copenhagan: GEC GAD Publishers, 1979.

Stinson, Karl W., *Diesel Engineering Handbook, 12th Edition,* Stamford: Business Journals Inc., 1972.

Woodward, John B., *Low Speed Marine Diesel,* New York: John Wiley & Sons Inc., 1981

Index

D

E

F

G

H

I

J

S

T

U

V

W

More Titles from Cornell Maritime Press

American Merchant Seaman's Manual
William B. Hayler, Editor in Chief
978-0-87033-549-5

Applied Naval Architecture
Robert B. Zubaly
978-0-87033-475-7

Behavior and Handling of Ships
Henry H. Hooyer
978-0-87033-306-4

Business of Shipping
Lane C. Kendall and James J. Buckley
978-0-87033-526-6

The Cornell Manual
John M. Keever,
978-0-87033-559-4

Diesel Engines
Leo Block, P.E
978-0-87033-418-4

Formulae for the Mariner
Richard M. Plant
978-0-87033-361-3

Handbook of Rights and Concerns for Mariners
Roberto Tiangco and Russ Jackson
978-0-87033-530-3

Marine Cargo Operations
Robert J. Meurn and Charles L. Sauerbier
978-0-87033-550-1

Marine Radionavigation and Communications
Jeffrey W. Monroe and Thomas L. Bushy
978-0-87033-510-5

Marine Refrigeration and Air-Conditioning
James A. Harbach
978-0-87033-565-5

Master's Handbook on Ship's Business
Tuuli Anna Messer
978-0-87033-531-0

Modern Marine Engineer's Manual, Volume I
Everett C. Hunt, Editor in Chief
978-0-87033-496-2

Modern Marine Engineer's Manual, Volume II
Everett C. Hunt, Editor in Chief
978-0-87033-537-2

Modern Towing
John S. Blank, III
978-0-87033-372-9

Nautical Rules of the Road
B. A. Farnsworth and Larry C. Young
978-0-87033-408-5

Primer of Towing
George H. Reid
978-0-87033-563-1

Real Time Method of Radar Plotting
Max H. Carpenter and Wayne M. Waldo
978-0-87033-204-3

Shiphandling for the Mariner
Daniel H. MacElrevey and Daniel E. MacElrevey
978-0-87033-558-7

Shiphandling with Tugs
George G. Reid
978-0-87033-354-5
Stability and Trim for the Ship's Officer
William E. George
978-0-87033-564-8

Survival Guide for the Mariner, Second Edition
Robert J. Meurn
978-0-87033-573-0

Tanker Operations
Mark Huber
978-0-87033-528-0

Vessel Security Officer
Joseph Ahlstron
978-0-87033-570-9

Vessel Traffic Systems
Charles W. Koburger, Jr.
978-0-87033-360-6

Watchstanding Guide for the Merchant Officer
Robert J. Meurn
978-0-87033-409-2

ISBN-13: 978-0-87033-590-7
53500
9780870335907